Renewable Energy
Strategies for Europe

Volume II
Electricity Systems and
Primary Electricity Sources

First published in Great Britain in 1997 by
Royal Institute of International Affairs, 10 St James's Square, London SW1Y 4LE
(Charity Registration No. 208 223) and Earthscan Publications Ltd, 120 Pentonville Road,
London N1 9JN

Distributed in North America by
The Brookings Institution, 1775 Massachusetts Avenue NW,
Washington DC 20036-2188

A catalogue record for this book is available from the British Library.

ISBN 1 85383 284 7

Earthscan Publications Limited is an editorially independent subsidiary of Kogan Page
Limited and publishes in association with the International Institute of Environment and
Development and the World Wide Fund For Nature.

Typeset by Koinonia, Manchester
Printed and bound by Biddles Limited, Guildford and King's Lynn.
Cover by Visible Edge.
Cover illustration by Andy Lovel.

To Walt and Cleonie Patterson

Contents

Tables

Figures

Boxes

Preface and Acknowledgements

This is the second volume in the series *Renewable Energy Strategies for Europe*. This volume has been long in gestation, but fortuitously it reached its final stages just as European governments, after eight tortuous years of negotiation, completed the Directive on Common Rules for the Internal Market in Electricity – more commonly known as the 'Liberalization Directive' – that has set the stage for progressive introduction of competition into European electricity systems. Such liberalization has emerged as a key theme in this volume; somewhat to my surprise, the research has led to the conclusion that liberalization is a precondition for Europe to exploit the full potential of its primary renewable electricity resources. Though the progress of renewables will depend on many other aspects of government policy as well, a rapid expansion of renewable electricity contributions in Europe now seems more plausible than it did when the study began.

The study has benefited from many useful inputs. I am indebted to my co-author Roberto Vigotti for contributing drafts of Chapters 2 and 5, including much material on the structure of power system and photovoltaic system costs that would otherwise have been inaccessible, and for making several revisions as the schedule of this project slipped. Michael Bishop conducted background research which helped to make my suggestions on North Sea wind energy a little less speculative than they otherwise would have been.

Many people read the draft text and offered comments which helped to shape the final product. Rodney Janssen, Catherine Mitchell, John Twidell, Christophe de Gouvello, Nicola Steen and Dominique Finon made substantive comments covering much of the text; Kjell Roland, Stephen Salter, Mike Allington, Peter Fraenkel, David Milborrow, Robert Hill, David Thorpe and Les Duckers added detailed knowledge on specific chapters. To them all I am most grateful.

The series of reports is based on a study initially sponsored by the

Research Directorate of the European Commission, with additional subsequent support from the UK's Energy Technology Support Unit; to Wolfgang Palz and Godfrey Bevan, in particular, I am most grateful for helping to arrange this. Most of all, I am indebted to my colleagues at the Energy and Environmental Programme, including Matthew Tickle, who has now moved on after six years with the Programme, Ben Coles, and Christiaan Vrolijk who has provided unstinting and invaluable help on the project since his arrival with us as a research assistant in January 1997. Margaret May and Harriet Ogborn of RIIA's Publications Department displayed their usual skill and tolerance in dealing with last-minute changes.

In some areas the study has drawn upon experience from research for my PhD thesis, which concerned the long-term integration and assessment of intermittent power sources such as wind energy, and upon my subsequent work in power system planning at the Department of Electrical Engineering at Imperial College, London. Working on this volume has on several occasions brought a sense of returning to my research roots, after almost a decade's experience of research on broader energy-environmental economics and policy, and it has reminded me of the debt I owe to those who helped to guide me in my earlier work. Particular thanks are due to Nigel Evans of Caminus Energy; and to Dillon Farmer of Imperial College, now deceased and much missed.

In understanding and applying insights relating to the momentous changes that are now shaking the electricity industry – changes that were almost inconceivable even a decade ago – the study has benefited greatly from the insights generated by another research project in the Energy and Environmental Programme, namely Walt Patterson's work on 'Transforming Electricity'. Walt has spent much of his professional life as a lively, provocative but widely respected critic of 'conventional wisdom' in the electricity business, and he has proved a valuable and cheerful presence on our Programme. He and his wife have also been personally supportive during difficult times, despite Walt's own health concerns; and to them this volume is dedicated.

March 1997   Michael Grubb

About the Authors

Michael Grubb is Head of the Energy and Environmental Programme at the Royal Institute of International Affairs, where he is responsible for directing a wide range of research on international energy and environmental issues. He is well known for his work on the policy implications of climate change, and other publications include a book on Emerging Energy Technologies, and one examining the outcome and implications of the Rio 'Earth Summit' UN Conference on Environment and Development. In addition, he is an adviser to a number of international organizations and studies, particularly concerning economic and policy aspects of climate change, and was a lead author for the Intergovernmental Panel on Climate Change.

Dr Grubb graduated in Natural Sciences from Cambridge University and gained his PhD, on electricity systems and renewable energy, at the Cavendish Laboratory in Cambridge. Before joining the RIIA he worked at the Cavendish and at Imperial College in London, where his studies resulted in various publications on the planning of electricity systems and the economic prospects for renewable energy sources.

Roberto Vigotti is responsible for renewable energy programmes at the Italian State Electricity Company ENEL. He is widely known for his contributions on distributed utilities, and on applications of photovoltaic technology.

List of Abbreviations

AC	alternating current
ALTENER	Alternative energy programme of the European Commission
BOS	balance-of-system
CCGT	combined-cycle gas turbine
CEE	central and eastern Europe
CIS	copper-indium diselenide; Confederation of Independent States
DAB	Deutsche Ausgleichbank
DC	direct current
DNC	declared net capacity
EC	European Commission
EDF	Electricité de France
EEA	European Economic Area
ENEL	Italian State Electricity Company
EPRI	Electric Power Research Institute US
ESHA	European Small Hydro Association
ETSU	Energy Technology Support Unit, Harwell
IEA	International Energy Agency
JOULE	Joint Opportunities for Unconventional or Long-term Energy Supply
NFFO	Non-Fossil Fuel Obligation
PV	photovoltaic(s)
RPS	Renewable Portfolio Standard
RTD	research and technology development
T&D	transmission and distribution
TERES	The European Renewable Energy Study
THERMIE	European Energy Technology Demonstration Programme
We	watts electric output
Wp	peak wattage output from PV

Summary and Conclusions

Technologies for generating electricity from primary renewable sources – hydro, wind, photovoltaic (PV) solar cells, wave and tidal power – are developing rapidly. So too are Europe's electricity systems, which are undergoing momentous change as governments and the European Commission move to liberalize the sector and to strengthen environmental and related policies. Despite Europe's high population density, these combined trends mean that renewable sources (including large hydro) are likely to supply between a quarter and a half of EU electricity by 2030. Appropriate policies would lead towards the upper end of this range (with continuing expansion of PV and offshore renewables thereafter) and could give Europe a lasting lead in some of the major industries of the twenty-first century.

Value and differentiation

Mandated premium payments for electricity generated by renewable sources have enabled the nascent renewables to emerge as commercial industries during the 1990s, but the present policies cannot be the main basis of long-term expansion. Wider development of premium applications and resources can provide the revenue streams and market experience required to establish commercially mature industries by realizing economies of scope and production scale. This will require full exploitation of the distinct features that characterize the primary renewables – lack of pollutant emissions, dispersed resources, capital intensity, variability of output, and (excepting large hydro and tidal schemes) small unit size and quick construction.

The small scale of most primary renewables and their lack of dependence on water or gas infrastructure gives them a comparative advantage in supplying island and other isolated demands; such applications are limited to 2–3% of the Union's population. For grid supplies also, system conditions vary widely across Europe in terms of fuel mix, capacity needs, environmental conditions, distribution costs and losses, and electricity costs and prices. Domestic tariffs are at least twice the cost of generation, sometimes reaching over 15∈/kWh, and the cost of grid supply to remote, island or mountainous regions that comprise about 10% of the Union's population is much higher than the system average; the same can be true for dispersed rural supplies. Connecting primary renewables directly to distribution systems to meet local demands may typically add 5–20% to the value of the electricity as compared with centralized generation; the premium could be much higher for applications with local storage, in more remote locations, or for PV connected at the point of end-use or at weak points in networks in southern Europe. Short construction periods may also yield significant benefits of flexibility given uncertain prospects for electricity demand. The development of better technologies for monitoring, control, load management and small-scale storage opens the possibility of realizing such potential benefits through 'distributed resource' electricity systems.

For most grid-connected applications except PV, the variability of output from primary renewables in Europe is not a significant technical or economic drawback. The seasonal correlations between electricity demand and the availability of wind, wave and hydro power in most of Europe increase their value relative to equivalent constant output. For system penetration up to 5–10% from each source this is economically more important than their peak availability, making electricity from these sources more valuable than from conventional sources; short-term fluctuations from dispersed applications also are negligible compared to those of electricity demand. Variability does, however, limit the scope of distributed benefits. The situation for PV is more complex, but in certain applications (including mounting on some service-sector buildings) its value may be substantially enhanced by its variability.

All these factors illustrate that the *value* of electricity differs according to its application and source. In addition to the six *external* issues (mostly benefits) noted in Volume I, six internal *systemic* issues can be identified associated with the dispersed, small-scale, modular, time-dependent and expanding market characteristics of primary renewable electricity technologies; these are also mostly beneficial to renewables. Correspondingly, a key issue is the *differentiation* of electricity as a product:

- *Geographical differentiation* rewards dispersed sources that supply local loads or bring economic activity to more depressed regions, while concentrated and often foreign or offshore sources supply urban and industrial demands: simultaneous localization and internationalization of electricity supply are not contradictory trends, but reflect better matching of resources to applications and also reflect the parallel development of sub-national and European-level political structures;
- *Temporal differentiation* rewards sources that supply more power in periods of greater electricity demand;
- *Environmental differentiation* rewards sources that have lower impact on local, regional and global environment, as reflected not only in prices but also in consumer preferences and local planning procedures;
- *Risk differentiation* rewards sources that can be purchased as guaranteed standardized products that are small and quick to install and that reduce dependence upon uncertain fuel markets.

Liberalization is the most plausible route to realizing such differentiation.

Characteristics and contributions

Hydro power, predominantly from large dams, currently supplies about 14% of electricity in the EU. Although economic potential exists to expand the contribution of large hydro by up to 40%, only a fraction of this will be exploited because of environmental concerns; also liberalization will discourage large schemes within Europe. Small hydro schemes face much less opposition and would benefit from liberalization; their contribution (including refurbishments and retrofits) could expand from about 1% to 3–5% of European electricity. Liberalization may increase the scope for investment in hydro power outside the EU for power exports; untapped potential in Norway, some parts of the former Soviet Union (notably western Russia and Georgia) and Iceland could supply several per cent of EU electricity.

Wind energy capacity has expanded by 25–30%/yr during the 1990s and its contribution will approach 1% of European electricity supply by 2000. Small private companies have led this explosive growth, and although liberalization will threaten the systems of premium payments on which they have depended, it will also remove important institutional obstacles that would have impeded further growth. Visual impact, and the costs in less windy areas, are the primary constraints, and the likely stable long-run contribution is 5-10% of European electricity supply from land-based and coastal windfarms. Most of this capacity could be installed in the next 2–3 decades, making important contributions to environmental targets on these timescales.

Long-run generation costs from both small hydro and wind energy are likely to be in the range 2.5–5.5∈/kWh depending upon siting and interest rates. For supplying bulk power markets (in addition to distributed applications), this could make these sources potentially competitive with clean coal or nuclear plants at many sites, but only at the very best sites could they compete unaided against CCGTs.

Solar photovoltaic technology is currently much too costly to contribute significantly to European electricity supply. The small current market implies considerable scope for cost reductions from scaling up production, and many technological improvements are possible, but it is implausible that PV in most of Europe can compete against centralized power production for the foreseeable future. However, its unique modular characteristics imply wide scope for 'niche' applications. 'PV cladding' and rooftop systems for meeting electricity demands in buildings, perhaps with a small amount of storage and/or feeding into local DC networks, could plausibly meet several per cent of European electricity demand; grid reinforcement is another important niche. PV could in principle have a comparative advantage over centralized power production for several other sectors, including electric transport (marine and land-based), though complete systems require much further development. Solar thermal electricity could develop as an important source in north Africa, in the long term potentially with exports to the EU.[1]

Tidal energy from large dams is technically easy but unlikely to be developed because of environmental opposition, the dominance of nuclear power in France and the economic impact of liberalization elsewhere. Small tidal dams are more feasible, as an element in local economic and amenity development, and could also form foundations for wind and wave energy

---

[1] Solar thermal power is not strictly speaking a primary electricity source, and its dependence upon direct sunlight and economies of scale make the prospects for generation in most of Europe poor; it is considered here only in the context of north African development.

devices. Extracting energy from tidal streams is a potentially important but little-explored option for large-scale exploitation.

Offshore wave and wind power technologies could in principle each contribute 5–10% of EU-15 electricity, but their costs are very uncertain because of the lack of a sustained research and development community; still less is known about the economic scope for tapping tidal stream energy. Coastal tidal and wave technologies form a resource that is negligible on a European scale, but that could be significant in local (e.g. island) supplies, and such applications form a useful bridge to developing technologies for exploiting the much larger offshore resources.

For most of these technologies, the potential in Europe is dwarfed by the global potential. European companies dominate international trade in small hydro and wind energy, but they are not strong financially in terms of promoting exports. Protection of domestic manufacturers has outlived its usefulness; open procurement, with greater industrial consolidation, will strengthen the ability of Europe's renewable energy companies to compete internationally. Wind energy is particularly well placed to become a major European export industry. PV is a more competitive international market and European comparative advantage is most likely to emerge in advanced applications, such as cladding and DC systems for buildings.

Geographical dimensions

In addition to off-grid supplies and PV-building and grid-support applications, three distinct geographical classes of niche markets combine above-average electricity costs with good and varied local primary renewables: islands, coastal areas, and mountainous regions. Technology developments including advanced load management and control systems have improved the prospects for combining different renewables locally to diversify the output and benefit jointly from connection to storage and other complementary infrastructure and generation technologies (which may include electricity from biomass, considered in Volume III).

Development of such geographic niche markets is inhibited by a lack of relevant institutions in such regions; the fragmentation of the different renewable energy industries; the prevalence of uniform electricity pricing;

and the institutional structure of current electricity systems, which inhibits small-scale developments that depend upon local resources. Liberalization could address the last two obstacles but its overall impact depends on the form it takes. Politically, the most difficult step – but perhaps the most crucial one for renewables – will be to encourage differentiated, cost-reflective electricity pricing.

It is harder for dispersed primary renewables to compete in supplying concentrated urban and industrial demands, though wind energy will do so to an important degree. However, developments in long-distance DC transmission technology improve the prospects for tapping large concentrations of renewables around Europe's periphery, which would supply such demands. To the south these may include solar thermal power, hydro and wind power from north Africa, though such developments would initially be focused upon suppling burgeoning local demand.

In northern Europe, options include Norwegian, Scottish and Icelandic hydro, wind and geothermal resources, and offshore renewables. The costs of dedicated transmission and foundations or moorings are likely to render isolated offshore generation impractical. Significant development is most likely in the context of integrated offshore systems, initially in the North Sea, drawing upon decommissioned oil and gas facilities for foundations or moorings, and upon sub-sea DC transmission cables which may also be used to export surplus hydro and wind energy from Norway and perhaps Scotland to the Netherlands and Germany. Development of Icelandic renewable resources for export via sub-sea cables could open up the North Atlantic wave and oceanic stream resources.

Such offshore developments are speculative but at present cannot be rejected on the basis of engineering cost estimates. The main need is to develop a stable research and development community including existing marine industries, leading to criteria and candidates for prototypes. Supplying power for offshore activities (e.g. gas pumping) could provide an initial entry market, offset costs, test integration with existing offshore infrastructure, and draw in the relevant existing offshore industries. However, it is unlikely that offshore renewables or most imports could compete unaided against CCTGs without big gas price rises; implementation is thus partly an issue of diversifying long-term import dependence.

Liberalization and policy

European electricity systems are undergoing profound technological and regulatory changes. The need to retire old plant and meet new demand growth may require as much new capacity to be installed in the period 1995–2020 as currently exists, with total investments exceeding 500 billion ecus; there is a wider choice of options than ever and the nature of such investments will depend upon fuel markets, technical advance, environmental and related policies, and regulatory and planning structures.

Outside France, there is an inexorable trend towards introducing competition in electricity supply. Although it will take different forms in different countries, liberalization will both undermine existing regimes for premium prices and open new opportunities. The heavily-regulated, fully competitive UK model will allow companies specializing in 'green power' and energy services to supply consumers directly. The less formalized systems arising in much of continental Europe will give greater scope for governmental influence to promote renewable investments by private companies. In some cases, electricity distribution companies may compete with centralized generators by promoting distributed generation including renewables.

Liberalization is the most plausible route to harnessing innovation and private finance, differentiating electricity sources and applications, and bringing other existing expertise and knowledge of local resources to electricity production. Liberalization is a *necessary* condition for fully utilizing the primary renewables, but is not a *sufficient* one, because of its impact on information, investment time horizons and reflection of external costs and benefits. Most specifically, although CCGTs and primary renewables are in most respects highly complementary sources, liberalization of both electricity and gas networks could lead to investments in CCGTs eclipsing all other sources including renewables.

Current systems for premium payments in continental Europe cannot be sustained in the face of growing renewable capacities and growing competition, and the UK's non-fossil fuel obligation also has important drawbacks. A system requiring electricity supply companies to obtain credits equivalent to a 'standard portfolio' of renewables would best stimulate least-cost applications, minimize adverse competitive impacts, and allow internationalization of support systems. Such a system could define separate bands for different technologies and be introduced by member states, as appropriate to national circumstances, in the context of EU negotiations on implementing climate change commitments. However, the longer such a system is delayed, the more difficult it may become owing to rising differential impacts in an increasingly competitive electricity industry.

Although the principal justification for such a tool is supporting the industrial transition from RTD to mature commercial industries, the political difficulty of reflecting environmental and other external costs directly in the electricity sector means that such a credit system may also be sustained in the longer term as a way of reflecting external benefits, perhaps unbanded and in combination with other mechanisms for supporting the commercialization of less developed technologies.

Wind energy in particular needs to be integrated in local and regional development plans and planning systems, to encourage sensitive siting and minimize local environmental disturbance and hence opposition. At the EU level, it is important to incorporate consideration of renewables into the Union's financial instruments, notably Structural Fund expenditures especially in the cohesion and future accession countries, expenditure by the European Investment Bank in eastern Europe and the Maghreb, and, particularly for PV, applications of development aid under the Lomé Convention.

Projections by the European Commission and related modelling studies underestimate the likely contribution of primary renewables because they do not reflect the structural and institutional changes that would accompany growing capacities. Within three decades, renewable sources could contribute as much as half of Europe's electricity. However, the rate and degree of development will ultimately hinge upon the extent to which governments accept liberalization as a means to, but not the end of, achieving intrinsically more efficient, diverse, and sustainable electricity systems in Europe.

Chapter 1

Electricity Systems in Europe: Economics and Emerging Structures

Many of the most promising renewable energy sources produce electricity. The primary renewable electricity sources of wind, solar electricity, hydro, wave and tidal energy – here called the 'primary renewables' – share characteristics of capital intensity, negligible atmospheric emissions, and variability of output; many are also inherently small in unit scale. The prospects for these sources depend heavily upon the characteristics of the rest of the system.

The bulk European power systems are technically relatively mature, but the power market is far from saturated. The European Commission's Energy Directorate projects that in the EU-15, almost as much new capacity will be constructed in the period 1995–2020 as currently exists, with associated total investment in the range 500–700 billion ecus.

There are major regional variations. In the UK, France and the Benelux countries electricity demand growth is slow, and coal and oil generation have been largely displaced by nuclear and increasingly gas-fired genera- tion, with excess capacity overall. This forms difficult competitive genera- tion conditions unless a gas shock or nuclear accident reverses these trends. In southern Europe and Ireland, demand growth is much stronger, nuclear power is improbable, and relatively inefficient coal and oil plants are likely to remain mainstays of generation for many years; in principle these are much more promising conditions. In Germany and Denmark the position is more ambiguous, with slower demand growth but resistance to nuclear power and reluctance to move too far or fast towards gas generation, while conditions in Scandinavia may be dominated by Sweden's planned nuclear phase-out.

Investment in renewables will increasingly depend upon its profitability for the investor relative to other options. The economics of electricity have *conventionally been assessed in terms of cost per unit of electricity output. However, the assessed costs even of conventional plant vary considerably according to national circumstances and economic assumptions; histori-cally estimated costs in the early 1990s ranged between 3 and 8 ∈/kWh. Electricity from new gas plant costs 2.5–3.5∈/kWh, and from advanced 'clean coal' costs in the range 4–5∈/kWh, depending particularly upon fuel prices and interest rates. The European 'carbon tax' at the level originally proposed, or technologies for removing CO_2 from power plant emissions, would add 20–50% to these costs.*

The costs of large-scale hydro power span the range of conventional generation costs, as is the case for wind energy in some locations already, and more widely in projections. Other primary renewables remain more expensive, but may be as cheap as conventional options for small-scale decentralized applications. Simple unit-cost comparisons would thus suggest that there will be some investment in hydro and wind energy at good locations, but, except for some decentralized applications, not in other pri-mary renewables. 'Internalization' of atmospheric pollution costs alone, at existing or proposed levels, would widen applications but not fundamen-tally change this picture.

However, electricity tariffs in Europe substantially exceed generation costs and vary widely between sectors and countries, indicating far greater complexity. Excluding outliers, industrial tariffs range from 5–7 ∈/kWh and domestic tariffs are typically twice this; in Germany, Denmark and Spain they exceed 15 ∈/kWh. The projected costs of most renewables fall within or below the range of domestic tariffs. This points to the importance of examining full costs, and the institutional structures that determine how these are translated into prices.

The institutional structure of electricity in Europe is at the beginning of momentous change. A century's development based upon expansion of scope and generation scale and organized through integrated monopoly utilities is giving way to more flexible, 'unbundled' and sometimes competi-tive structures. The Liberalization Directive of December 1996 reflects the inexorable march of liberalization of European systems, at least outside France. Though it will take different forms in different countries, this will fundamentally affect the prospects for electricity investments everywhere.

Competitive structures raise the costs of capital and increase competitive pressures from other new sources, such as advanced gas plant. In general liberalization will raise perceived costs but may lower long-run tariffs, and it will discourage coal subsidies and investment in conventional thermal plant, while promoting generation from natural gas. The short-run impact on the economics of renewables falls between these two.

Currently, most governments support renewables electricity investment through systems that include premium payments for such electricity, often related to tariffs, at levels mostly in the range 3–10 €/kWh. These systems have been at the root of the rapid growth of renewable electricity in Europe in the 1990s. Such direct supports will come under increasing pressure from liberalization, which in exchange, and in combination with continuing technical advance and policy developments, will open hitherto inconceivable opportunities for renewables.

1.1 Introduction

Electricity production accounts for about one-third of primary energy requirements and greenhouse gas emissions in Europe. Many of the most promising renewable energy sources produce electricity. This volume examines European electricity systems and the renewable sources that generate 'primary electricity' – the technologies such as wind turbines, solar cells, and various water-based technologies that generate electricity directly from renewable resources. The next volume in this series examines other renewable sources, including those like biomass and geothermal energy, that can be used to generate electricity by conventional means using heat obtained from the renewable fuel.

The deployment of these sources may depend heavily upon their value to the electricity system. This depends not only upon their costs and characteristics, but also upon the underlying structure and pricing systems governing electricity supply, which, together with the wider market for power generation, affect the choice between different power sources. Since the power system itself is so crucial to any realistic appraisal of the prospects for renewable electricity sources, this chapter and the next one focus upon the system issues; the following chapters consider specific technologies.

Electric utilities in Europe – and most of the world – are structured around large, central power stations, connected to transmission systems which deliver the electricity to the customers on distribution networks. The output from these power stations is controlled – 'dispatched' – so that the stations produce in order of increasing cost (short-run marginal cost) as the demand rises. This structure started developing when small generating companies connected their lines with one another, forming the beginnings of vast networks. The coordination of generation with transmission and distribution then occurred within a single vertically integrated system. This feature, eventually coupled with monopoly franchise, was deemed necessary to ensure integrated operation, and to allow economies of scale by protecting the long-term investment in these large facilities. In return, these monopoly utilities were subject to government oversight and regulation, with an obligation to plan and implement investments to ensure adequate and stable supplies. Such centralized and integrated power systems, with the power delivered (and sometimes generated) by monopoly distributors, became the dominant pattern of electricity system development around the world. The system planners decided what capacity should be built – and almost universally they selected large, central power stations.

In the past ten years, this pattern has begun to break down. The economic basis on which decisions were taken have been shaken by altered trends in demand, input costs, technology developments and environmental pressures; and regulatory structures are changing to allow new entrants and new decision-makers using different criteria. The whole context for decision-making concerning power system investments is changing, in ways that have profound implications for renewable energy. This chapter examines these changes.

1.2 European electricity technologies and systems: fuels, costs and prices

Costs of conventional generating technologies

The economic context for renewable energy investments is set in part by the costs and prices on existing electricity systems, and the cost of alternative options for generation. A natural starting point might thus seem to be a comparison of generating costs. Until 1992, the International Energy Agency

Table 1.1 Estimated costs of different power generation technologies

Country	Lower discount rate (5%)			Higher discount rate (10%)		
	Nuclear	Coal	Gas	Nuclear	Coal	Gas
UK	4.84–5.16	4.88–5.16	4.52	7.74–8.06	5.81–6.61	4.68
	[4.41–4.70]	[4.45–4.70]	[4.12]	[7.05–7.35]	[5.30–6.02]	[4.27]
Germany	5.31	8.01–8.74	–	7.74	8.01–9.36	
	[4.84]	[7.30–7.97]	–	[7.05]	[7.30–8.53]	–
France	3.28	5.08	5.48	4.52	5.89	5.79
	[2.99]	[4.63]	[4.99]	[4.12]	[5.37]	[5.28]
Denmark	–	3.50	3.50–3.72	–	4.41	4.13–4.32
		[3.19]	[3.19–3.39]	–	[4.02]	[3.76–3.94]

Note: Costs are in 1991 US dollar cents (IEA) and [1995 ecucents] per kWh; prices in 1995 ecucents have been calculated using the consumer price index for Europe for inflation correction.
Source: IEA, *Projected Costs of Generating Electricity, Update 1992* (Paris, 1993).

(IEA) regularly published estimates of generating costs; their most recent published estimates are shown in Table 1.1.

The established technologies for coal and nuclear plant in this table are quite mature, and fuel prices (which do not anyway represent a high fraction of the costs) have not changed greatly since the early 1990s, so that if other factors are similar the costs given for coal and nuclear plant remain relevant; newer technologies are considered below. The table also indicates some important qualitative features. Notably, the costs and indeed ranking of options differ considerably between different countries, and are sensitive to the discount rate, with gas generation becoming relatively cheaper at higher discount rates.

This points to the sensitivity of generating cost estimates to national economic and other circumstances. Until the late 1980s, electricity generating investments were made under fairly stable and comparable conditions in most countries; centralized utilities invested in new plant to meet projected demand growth under economic criteria established by regulatory authorities, notably using 'test discount rates' of 4–8% with almost unlimited ability to pass unexpected costs through to consumers. There was also considerable stability in the basic technologies of thermal power stations: whether the fuel was coal, oil or nuclear power, these plants (along with hydro) involved capital investments exceeding 1000 ecus/kW (1990 money) in units of several hundred megawatts, and generally taking at least five years to construct, often longer.

By the early 1990s all these simple certainties were disappearing. The more erratic course of electricity demand injected new uncertainties about the role of new plant in the system and when it would be required. New, independent generators were increasingly given access on a competitive basis, and some systems were privatized, bringing in new actors with different economic criteria, notably higher discount rates, shorter time horizons and far greater sensitivity to investment risks. The arrival of combined-cycle gas turbine plants (CCGTs), together with continually increasing estimates of natural gas reserves (see Volume I,[1] Chapter 2), highlighted a new dimension of technological uncertainty, as unit sizes fell and capital costs and construction timescales plummeted, but with varying and often contradictory perceptions about the prospects for gas prices. The rise of environmental concerns added a new dimension, as the underlying cost structure was increasingly eclipsed by the regulatory requirements for meeting national standards and obtaining planning permission.

Thus the costs of generation have increasingly become functions of public policy. It is symptomatic that in the early 1990s the big four countries of the European Union could each give a different account of its cheapest source: coal in Germany (monopoly utilities in electricity and gas, and with subsidized coal production); nuclear power in France (large-scale standardized production developed by the national utility with unfaltering governmental backing); oil in Italy (with coal and nuclear energy being publicly unacceptable and institutionally unmanageable); and gas combined-cycle plants taking everything by storm in the UK's liberalized markets. The malleability of electricity economics, and its dependence on the values and perceptions of the investor and system, is a theme that emerges in different guises throughout this volume. Cost comparisons are not irrelevant, but (unless the gap between different options is really very large) they need to be placed carefully in the context of specific investment conditions.

---

[1] M. Grubb, *Renewable Energy Strategies for Europe, Volume 1: Foundations and Context* (London: RIIA/Earthscan, 1995).

Impact of advanced technologies and environmental constraints

Technologies for fossil-fuel generation have also developed, partly in response to changing economic and environmental conditions. The rapid emergence of CCGTs has been accompanied by considerable reductions in their cost, so that gas power generation in many locations is now much cheaper than indicated in Table 1.1. In addition, the costs of removing SO_2 and NO_x from coal-powered stations has declined, and various advanced 'clean coal' technologies have been demonstrated. Desk-based studies have also produced estimates for the costs of removing CO_2 emissions from coal and gas power generation.

Table 1.2 shows estimates of costs from such advanced fossil-fuel options, which may provide a more relevant indication of the technologies against which renewables would be competing, together with the impact of higher discount rates. Among other things, this table illustrates the reasons for the 'dash to gas' in several liberalizing power systems. Modern CCGTs offer very low-cost power. Gas prices in most of Europe are between price levels A and B in Table 1.2, and yield prices of 2.5–3.5 ∈/kWh at private sector discount rates; relevant costs for 'clean coal' technologies are more in the range of 4–5 ∈/kWh. It would take a severe and sustained gas price increase to make CCGT costs comparable with coal. Furthermore the investment cost per unit is about half of those of coal-fired plants, so that the higher discount rates which arise in a competitive environment for electricity production also favour gas-powered plants.

The table also illustrates the potential impact on costs of CO_2 constraints. The proposed European carbon tax, if implemented at the full level originally proposed in 1991 – which has so far proved too politically contentious to implement in the EU – would have added about 1.4 ∈/kWh for coal and half of that for gas. Furthermore, these estimates suggest that even CO_2 capture and disposal, which if accurate would represent an upper limit to the costs of CO_2 constraints, may not push the generation costs above about 5 ∈/kWh on central fuel price projections (or 5.5 ∈/kWh at higher interst rates). This gives a measure of the centralized generation costs against which renewable energy may have to compete, even when environmental costs including CO_2 abatement are factored in.

Forecasts for the costs of electricity from centralized and decentralized renewable energy technologies are shown in Figure 1.1 (a) and (b). The bars

Table 1.2 Costs of advanced fossil-fuel generation technologies

Plant type	Coal			Natural gas
	Typical with de-SO$_x$ and de-NO$_x$	Supercritical with de-SO$_x$ and de-NO$_x$	IGCC[a]	Combined-cycle
Status	Conventional	Established technology	Demonstration	Established technology
Special investment cost[b] (\$/kW) [ecus/kW]	1300 [1169]	1740 [1565]	1800 [1619]	750 [675]
Cost of electricity[c] (US¢/kWh) [∈/kWh]				
Fuel price level A[d]	3.9 [3.5]	4.1 [3.7]	4.2 [3.8]	2.4 [2.2]
Fuel price level B[e]	4.6 [4.1]	4.7 [4.2]	4.8 [4.3]	3.8 [3.4]
Fuel price level C[f]	5.0 [4.5]	5.1 [4.6]	5.2 [4.7]	4.4 [4.0]
Extra cost at high discount rate[c]	0.7 [0.6]	1.0 [0.9]	1.0 [0.9]	0.4 [0.4]
CO$_2$ performance and associated costs				
Typical carbon emissions (gC/kWh)	230	200	220	110
Extra cost due to European carbon tax (∈/kWh)	1.4	1.2	1.4	0.7
Extra cost for CO$_2$ capture in favourable locations (∈/kWh)[g]			0.7	1.8

[a] Integrated Gasification Combined Cycle.
[b] For 500-MW net units in mainland northern Europe (overnight build).
[c] Low discount rate is 6%; high discount rate is 10%.
[d] Fuel price level A based on gas at US\$ 2.2 /GJ [2.0 ecus/GJ] and coal at US\$ 1.2 /GJ [1.1 ecus/GJ].
[e] Fuel price level B based on gas at US\$ 4.5 /GJ [4.0 ecus/GJ] and coal at US\$ 2.0 /GJ [1.8 ecus/GJ].
[f] Fuel price level C based on gas at US\$ 5.5 /GJ [4.9 ecus/GJ] and coal at US\$ 2.5 /GJ [2.2 ecus/GJ].
[g] All component parts of the technology are available, but have not been demonstrated at scale in this application. Costs of disposing of the captured CO$_2$ depend heavily upon location.
Sources: Intergovernmental Panel on Climate Change, *Climate Change 1995: Impacts and Response Options (Working Group II)*, Cambridge University Press, 1996, Ch. 19.

Figure 1.1 Forecasts for unit electricity costs for centralized electricity generation from renewable energy technologies
(a) Centralized generation

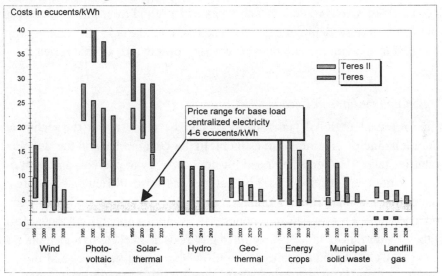

(b) Decentralized generation

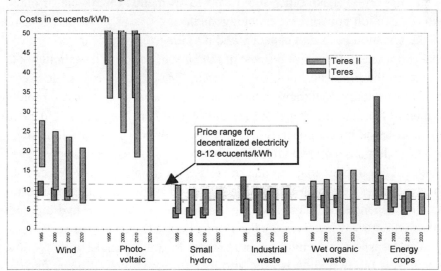

Source: *The European Renewable Energy Study II: The Prospects for Renewable Energy in 30 European Countries from 1995–2020*, EC DG-XVII (Brussels, Luxembourg: 1996).

indicate a crude range of possible costs, owing both to technological uncertainties and to differing resource characteristics. The figures compare these with the cost range for competing electricity, respectively centralized and decentralized. Cost projections and comparisons, and their implications for renewable energy deployment, are discussed in the specific technology chapters of this volume and the subsequent volume (III) of this series.

Regional variations in Europe

Some relevant regional generalizations can be made about the technical conditions facing electricity investment. In the UK, France and the Benelux countries there is no pressing need for new capacity at present, and dominance by nuclear and, increasingly, natural gas form difficult competitive conditions. Germany and Denmark remain heavily dependent upon coal, and alternatives are sought for economic and environmental reasons. Most of these countries have well-developed gas grids and access to abundant gas supplies. CCGTs would seem favoured in liberalized systems, and unless there is a significant gas price shock, competitive generating companies seem unlikely to make other significant investments, renewable or other, without explicit government encouragement.

The situation of excess capacity and dominance by nuclear, gas or even coal is not characteristic of the rest of the Union, nor does it seem likely to persist on the timescales of greater interest for potential large-scale use of renewable energy. Electricity demand has continued to grow more rapidly in much of southern Europe and Ireland, where many regions still depend upon oil-based power generation and seem likely to do so for a long time; Ireland still plans to expand highly carbon-intensive peat-based power generation. This implies that the marginal savings from renewable electricity sources would help both to displace higher-cost marginal fuels and to reduce oil dependence. In many of these countries, nuclear and hydro power are ruled out as alternatives, and coal may be relatively expensive because of both transport and environmental costs. Scandinavian electricity, now linked by a freely trading Nordic grid, is dominated by hydro and nuclear power, but if the planned phase-out of nuclear power in Sweden proceeds there will be a pressing need for replacement capacity.

Electricity tariffs in Europe

Some of these factors – and many others – are reflected in variations in electricity tariffs. These are displayed in Table 1.3. In most (but not all) countries electricity prices have risen somewhat during the 1990s, in part because of fiscal conditions (taxation and a move towards market-oriented loan rates and controls). There are three striking disparities in these data: the price disparity between small domestic and industrial consumers; the price disparity between different members of the Union; and the gulf between prices (especially for smaller consumers) and the estimated generating costs shown in Table 1.1 and Table 1.2. Prices to domestic users are typically twice the industrial tariff, sometimes more; and more than twice the estimated costs of generation illustrated in Table 1.1. Prices are generally highest in Germany, Denmark and the Iberian peninsula, and lowest in Sweden and Finland, Greece and (for industrial users) France.

All these variations are of considerable relevance to renewables. The gap between generation costs and delivered price (tariffs) involves complex issues associated with generating history, network and system costs, some of which are considered more fully in Chapter 2. The costs of many renewable sources already fall below or within the range of tariffs set out in Table 1.3, or with further development are expected to do so; they lie between existing conventional generating costs and existing sale prices. The questions of what payments renewables can or will attract, and how they should be related either to other generating costs or to sales tariffs, are therefore crucial to the economics of renewable energy investments. The answers depend partly on the role of intermittent sources such as wind and solar power in the system; this is considered in Chapter 2. But they also concern the organization of the system, with which this chapter is primarily concerned.

1.3 The evolution of electric utilities in Europe

From the 1900s to the 1970s, power plants were built in increasing sizes to capture the economies of scale. The strategy was to grow and build because new and larger plants had successively lower unit generating costs with fewer environmental constraints than today. Abundant, cheap fossil fuel and subsidized nuclear power encouraged this strategy. Consequently, today the

Table 1.3 Electricity tariffs (∈/kWh) in the European Union

	Household		Industrial	
	1991	1995	1991	1995
Austria	0.123	0.132[a]	0.053	0.065
Belgium	0.132	0.140[a]	0.049	0.047[a]
Denmark	**0.138**	**0.167**	0.052	0.055
Finland	*0.080*	*0.087*	0.049	0.050
France	0.112	0.133	*0.043*	*0.048*
Germany	0.127	**0.163**	0.070	0.080
Greece	*0.089*	*0.090*	0.052	*0.049*
Ireland	0.105	0.105	0.053	0.053
Italy	**0.138**	0.135	**0.084**	**0.074**
Luxembourg	0.095	0.116	n.a.	n.a.
Netherlands	0.091	0.108	*0.042*	0.056
Portugal	0.130	0.144	**0.102**	**0.097**
Spain	**0.158**	**0.155**	**0.082**	0.065
Sweden	*0.077*	*0.075*	*0.042*	*0.031*
UK	0.102	0.099	0.057	0.054

Notes: The table indicates prices paid net of rebates, i.e. net of Value Added Tax for industry, converted from US$/kWh at an exchange rate of $1=0.797 ecus.
Bold indicates highest three values; *italics* indicates lowest three values.
[a] Data for 1994 (1995 unavailable).
n.a.: data not available.
Source: IEA, *Energy Prices and Taxes* (Paris, IEA, 1996).

electric utility industry in Europe – east and west – is heavily dependent on coal, oil and nuclear energy for most of its generation. It is organized in utilities of a range of sizes, from small municipal units to the giant *Electricité de France*. But most except for the UK, Irish and island systems are highly interconnected, with synchronized operation and regular exchange of electricity.[2]

Until the 1970s, the decreasing cost of electricity to satisfied customers resulted in an indifference to how the utility conducted its business. Most

---

[2] In most cases this is a short-term exchange, but *Electricité de France* exports power, especially to the UK and Italy, and long-term trade across Europe – including the UK – is growing.

customers bought power with little concern about unit price and power quality. Likewise, the utility had very little reason to know how the customer used its product.

Towards the end of the 1960s, the limitations of growth based on these principles began to become apparent. Although it was the oil crises that most visibly precipitated the period of instability – which is still with us – the foundations of the system, built on increasing plant scales and centralization to reduce production costs, had already shown themselves to be shaky.

One sign that something was wrong was the halt in the trend of continually falling generation costs, which had levelled out and appeared to be heading up. Generating efficiencies, which had increased from a few per cent to around 40 per cent, stopped growing. This improvement had been achieved mostly by increasing steam temperature and pressures, but material limits had reached their ceilings and additional gains would have been possible only by introducing costly new materials or competing technologies.

The size of fossil-fuel and nuclear power stations peaked at 1400 MW. Boilers had become too big for the materials and construction techniques that were economically available; the more extreme conditions and the complexity of trying to squeeze additional performance made such plants less reliable, and the enormous size amplified the costs of such unreliability. Uncertainty in regulatory treatment of new big investments probably also contributed to this peak in unit size. It was clear at this stage that the 'economies of scale' in generation had reached or overshot their practical or financial limits and that, with the then available technologies, no additional cost compression was achievable, either in capital expenditure or in operating expenses.

It was against this turbulent backdrop that the oil crises of the 1970s broke. For the first time in many years, energy became a scarce and expensive commodity. Reduced economic growth, government action, and the reaction of consumers all combined to check energy consumption. Growth in electricity use – though still faster than that of total energy consumption – nevertheless was much slower than generally forecast. The golden rule of 'a two fold increase every decade' became obsolete as demand began to trace out a far more restrained and irregular trend.

Decisions on new plant capacity lagged behind these changes, but orders for new plants slowed considerably during the 1970s and the rate of construction and completion declined correspondingly during the 1980s. The construction industries contracted and merged, chasing ever fewer European orders and turning increasingly to overseas markets. The generating stock in Europe consists mostly of plants 10–40 years old; and for many utilities, the question of whether to maintain the older stock or replace it is beginning to loom large.

Along with the oil shocks, the environment also became a major issue in the 1970s. The effect was that it became especially difficult (if not impossible) to construct larger plants because their local impact attracted the hostility of residents near the chosen sites. The result was greater stringency in the bureaucratic procedures necessary to open work sites, carry out the work and operate the completed plants. The commissioning dates of plants under construction became highly uncertain; and the availability of permits and sites for new construction became ever more limited.

Environmental movements have forced many of the generators to 'clean up' their act. The resulting increased cost of electricity means that some of the environmental costs are now being internalized. These measures, however, do not remove all of the environmental externalities. Some countries now require or are considering measures to incorporate environmental costs more directly, as for example with taxes on SO_2 and NO_x emissions in Sweden.

The steady tightening of European policy on SO_2 and NO_x (see Volume I, Chapter 2) has particular continuing implications for electricity investments in some parts of Europe. In the countries which led the process, the internal response was simple: old plants were retrofitted with clean-up technologies to enable their continued operation, and the costs passed on to their captive consumers. A wide range of options is now available in other parts of Europe, however. Utilities are faced with the decision of whether or not to invest in existing ageing fossil fuel units in order to comply with the new standards being phased in up to 2010.

1.4 Implications for the scale of new investments in the EU

These various factors combine to determine the outlook for new generating investments, and they mean that despite the more modest growth of

Table 1.4 Projected electricity capacity requirements in Europe, 1993–2020

Source	'Conventional Wisdom' scenario		'Forum' scenario	
	New capacity (GWe)	Investment (billion ecus)[a]	New capacity (GWe)	Investment (billion ecus)[a]
Nuclear	42	71	120	239
Conventional thermal	116	179	79	132
New thermal	241	253	203	199
Renewable energy	53	69	96	118
Total	**456**	**571**	**502**	**689**

[a] Investments from 1991 to 2020 in 1995 ecus.
Source: European Energy to 2020, ECDG-XVII (Spring 1996).

electricity consumption there appears to be a considerable need for new capacity over the next few decades. The 'Conventional Wisdom' projection by the European Commission's Energy Directorate, summarized in Table 1.4, suggests that 456 GW of new capacity would be required out to 2020 – only a little less than the total installed capacity existing in 1990. The projection suggests that about one-third of the new investment would be made in combined-cycle gas turbines, and another quarter in conventional thermal power plants, with substantial other investment in new thermal technologies (such as 'clean coal'), co-generation and nuclear power. In total this represents investments of over 500 billion ecus over the period to 2020. Europe is not a small market for future power generation.[3] The new investments are also projected to be spread widely around the Union.[4]

However, primary renewable electricity sources are projected to account for little over 6% of new capacity out to 2020 (biomass, including power

[3] A different detailed consultants' study of the prospects for electricity investments worldwide over the period to 2020 also highlighted the likely scale of power investments in Europe due to continued demand growth, retirement of old plant, restructuring and environmental requirements. Six European countries were among the top 20 countries in terms of projected needs for new power generation, with projected requirements for over 300,000 MW of new generating capacity. (*The Future of the Electric Power Industry around the World*, DRI/ McGraw-Hill, 1996).

[4] The DRI study just cited indicates the top six countries in terms of capacity addition over 1996–2020 to be Germany (99 GW), UK (65 GW), Italy (47 GW), Spain (39 GW), Finland (13 GW) and Portugal (11 GW).

generation from wastes, and fuel cells are together projected to account for a similar amount), and the majority of this is from hydro power. The Commission also presented some other scenarios, one of which ('Forum') was projected to be more environmentally focused, with the renewable energy capacity addition increased by 60%. In this case, renewables in total (including biomass and fuel cells) attract investment of about 45 billion ecus (compared with 30 billion ecus in the 'Conventional Wisdom' scenario). In all these scenarios, however, the contribution of renewables is lower than called for in the Union's own ALTENER programme, or projected in the TERES studies (Volume I, Chapter 3 and this volume, Chapter 8).

Despite the adoption of different scenarios, the projections are based largely on a conventional conception of the structure of electricity systems and investment options, with the recent developments in CCGTs tacked on: essentially, they all project a continued expansion of centralized power generation in large conventional power stations plus CCGTs, with additional growth of co-generation and somewhat marginal contributions from renewables.

However, economies of scale in generation no longer dominate planning, and conditions seem set to change further. Not only have large thermal power stations reached their upper limit of scale, but new generation, transmission and system-control technologies have evolved, making possible the emergence of much smaller generation unit technologies able to duplicate or exceed the efficiencies of large ones. In several countries gas power generation has been the main initial beneficiary, and in all probability the projections are correct in assuming that this situation will continue. However, these new options include various renewable energy sources together with innovations in storage and customer energy efficiency measures, and the range of options open for power generation is expanding dramatically. This study explores in depth the contributions that renewable sources might make, and the conditions that might encourage more radical changes than are envisaged in the Commission's projections.

For key issues concern not just technologies and costs, but the whole structure and investment basis of the electricity industry. Most of the European utility industries remain staunchly conservative, and one other feature of the traditional structure, very relevant to renewables, is the lack of any real incentive to look at alternative, smaller-scale ways of generating power. For the top management of these huge companies, it was, and remains, far easier to consider one 1000 MW investment than 1000 investments of 1 MW, the typical reaction being, quite simply, that renewables are not big enough to bother about.

1.5 The march of electricity liberalization

This now seems set to change – along with much else in the European industries. The rising dissatisfaction with the traditional way of organizing the utilities, the disparities between electricity prices in different sectors and different parts of the Union, and the underlying thrust of EU economic policy, have together provided the impetus for the changes which are now shaking the foundations on which European electricity has been organized for a century.

These changes have big implications for renewable energy. As noted in Volume I, Chapter 7, regulatory change has played a central part in promoting renewables in the US especially, and more recently in the UK. In fact there are many different degrees and kinds of regulation that can be considered. The UK has gone furthest in promoting full competition in generation, and some central European utilities have moved towards integrated resource planning (IRP).

IRP can be, and in many US states has been, regulated so that the traditional planning process takes explicit account of environmental costs using 'externality adders'. These are monetized estimates, specified by the regulatory authorities, of the environmental impact of different options. The utility planners add these figures to the assessed financial cost of different options before deciding which represents the 'least cost' choice. This can apply both to the utility's own generation and to bids made to the utility by independent generators if the process involves competitive bidding. Because this tends to penalize fossil-fuel sources (especially coal), it improves the position of renewables. Sources can also be evaluated at a specific public-sector discount rate, which tends to favour renewables.

In Europe, however (and indeed now in the United States), regulatory reform has been moving in a different direction. In 1988, the European

Commission first signalled its ambition to extend the competitive principles of the Single European Act to the energy sector and, specifically, to try to introduce greater competition into Europe's gas and electricity industries. After very considerable opposition, in 1991 the European Council took a first step towards this with a Directive on 'unbundling' to introduce greater clarity into the cost structure of utilities.

Most electric utilities have strongly opposed moves towards liberalizing power systems. They fear it will erode their markets and perhaps destabilize the system, and argue that it will burden smaller customers with undue fixed charges or will strand investments.[5] Moves towards full open ('third party') access continued to be strongly opposed by most utilities and by some countries – most notably France, which sought to protect the national monopoly character of Electricité de France. But most governments did enact to allow for some degree of independent (non-utility) generation, and the pressures for allowing greater access for major consumers and for independent power generators on a competitive basis continued to grow.[6]

As the most dire warnings about the consequences of liberalization proved unfounded, more and more European governments took steps to open their systems to varying degrees and kinds of competition during the early 1990s. Emboldened by these developments, during 1995 the European Commission renewed its efforts and started to threaten legal action, through the European Courts, against countries that refused to open their systems to competition. Finally the European Council of Energy Ministers reached agreement in June 1996, and six months later passed the full 'Directive Concerning Common Rules for the Internal Market in Electricity', which entered into force in January 1997.[7] The agreement will start opening up the EU's electricity systems in stages over a period of six years from January 1997, to be followed three years later (in 2006) by a review of

---

[5] For a discussion see F. McGowan, *The Struggle for Power in Europe: Regulation and Competition in the European Electricity Industry* (London: Royal Institute of International Affairs, 1993).

[6] Ibid.

[7] 'Directive 96/92/EC of the European Parliament and of the Council, 19 December 1996, concerning common rules for the internal market in electricity' (OJ L 27/20, 30 January 1997).

progress. The Directive includes both opening up of competition for industrial consumers, and a requirement to give access to the system for independent generators on negotiated and transparent terms.[8] Despite all the limitations and caveats of the agreement, it represents a crucial step towards more open competition in European electricity and the most powerful sign yet that such trends are inexorable, driven by the pressures of those who want to buy and sell electricity on more competitive terms.[9]

The ultimate implications for power system investments will be huge. In a fully liberalized system anyone, in principle, will be able to generate electricity, for his/her own use and/or for selling directly to customers or to the system at published (often regulated) charges. Purchasers of electricity – whether large industrial consumers, distribution companies, or ultimately perhaps even domestic consumers (as planned for the UK in 1998) – will increasingly be able to choose their supplier, from anywhere in Europe. The concept of central planning for a nation's electricity would be radically altered if not gone.

The degree and form taken by legislation to inject more competition are likely to vary a great deal across the Union. Britain has already established a closely regulated competitive electricity market and will extend this to all final consumers in 1998. At the opposite end of the spectrum, France has ensured a high degree of continuing state control by invoking a 'single buyer' option in the Directive, as will Italy. German and Danish models of local monopoly suppliers will be retained at least for a long time, and it is likely that these and many other European countries will adopt a much less formal and heavily regulated system than the British, relying on negotiated

---

[8] For an analysis see *Financial Times, Power in Europe* (28 June 1996). The Directive requires countries to open a share of their electricity market equivalent to the share of industrial consumers over 40 GWh (by 1997), 20 GWh (by 2000), and 9 GWh (by 2003). Governments must also allow independent generators access to the system, either on terms of negotiated access or through the French 'single buyer' proposal which preserves monopoly control over system planning and network management.

[9] *Power in Europe* concluded: 'Admittedly the Directive is weak – limited in scope, weighed down with provisos and escape clauses, and stretched out over 9 tortuous years – but the significance of its birth should not be underestimated. As the UK's large consumer, ICI, said, the Directive will trigger competitive trends that will be hard to stop, forcing open Europe's heavily bolted power market doors.'

access backed by the threat of government intervention against unfair practices. Yet despite the different forms, it appears inevitable that competition will increasingly become the context in which policies towards renewable electricity sources will be placed; and – as illustrated later in this report – the way in which liberalization is implemented may in turn have big implications for renewable energy.

1.6 Implications of liberalization for the economics of electricity investments

In a fully competitive electricity system, in principle anyone could generate and sell electricity. A regulatory framework governs terms of access to the transmission network, together with planning procedures and compliance with other national laws (such as environmental legislation). Apart from this, the criteria which determine plant construction are financial viability and profit, on terms struck either directly with final customers or by selling the electricity into an electricity 'pool'.

This affects the economics of all electricity investments. In the UK, the particular structure of liberalization combined with the low price of natural gas and relatively short-term focus of British finance led to a 'dash for gas'; the liberalized companies have built nothing else except where given special incentives to do so, and there is no sign of this changing. In the rest of Europe, the structure of liberalization and the context are likely to be very different. But there are common themes, and various far-reaching implications for renewables.

Liberalization, in almost any form, will increase pressure on the German and Spanish governments to stop subsidizing domestic coal production. It will also tend to inhibit capital-intensive construction that has a long lead time – particularly construction of a plant involving other kinds of risks, such as nuclear power stations. Indeed, relative to new investments in any conventional thermal plant, the economics of many renewables should be improved. Furthermore, liberalization will tend to break open the utility structures that in practice have often retarded and opposed the development of vibrant and independent generating businesses. Liberalization gives freedom for innovation and exploitation of local advantages, especially for small-scale projects; there is no need to convince central utility authorities that the investment makes sense, since the entrepreneur can make the judgement and – if s/he can convince the banks – can take the risks. Liberalization should also make it much easier to trade power, for example from areas with a high concentration of renewable resources. In this sense, liberalization represents the greatest opportunities that renewable energy has yet seen.

Yet there are drawbacks. With liberalization, utilities can no longer act as agents of public policy. It becomes very difficult to reflect external costs other than by direct resource and pollution taxation, which is politically very difficult (witness the failure of the European carbon tax proposals), or by other, less tried mechanisms such as tradable emission permits. Research, development and demonstration activities plummet in the face of competition and the pursuit of short-term profits. Economic valuations are done at private-sector discount rates, which tell against capital-intensive options such as renewables, and (because of bankers' attitudes) can create an economic disincentive to application of less familiar technologies.

Indeed, the financial conditions created by liberalization appear most conducive to investment in natural-gas power stations, and if both electricity and gas systems are liberalized in Europe it is hard to see why banks would back any other kind of investment without explicit incentives for doing so.[10] The costs of gas power plant have continued to decline. The availability of gas in and around Europe has continued to increase. And, particularly if electricity liberalization is accompanied by liberalization of the gas industry, it is hard to see why many companies will build new capacity other than gas-fired (as CCGTs or for co-generation) as long as gas prices remain low – which could easily be a couple of decades. We revisit the overall economic characteristics of renewable electricity sources in liberalized systems in the final chapters of this book, and find that the outlook is not quite as difficult as is suggested here. Nevertheless, the dominance of

---

[10] This is not a universally accepted judgment, and may reflect excessively the UK experience; certainly, gas is not the cheapest option for power generation everywhere in Europe at present. Yet, in view of the massive availability of cheap gas for Europe over the coming decades and the attractive financial characteristics of the technology, it remains the authors' judgment that, if and as gas systems too are liberalized, the gas price will fall to a level that ensures its dominance of most new private-sector power system investments.

gas in new investments is the sobering context that could face renewable energy in much of Europe. And, at the same time, liberalization could further increase the pressure on the current regimes for supporting renewable energy.

1.7 Current policies for supporting renewable electricity generation in Europe

For the reasons sketched in Volume I of this series, many European countries have programmes to support renewable sources of electricity. Beyond explicit support for research and technology development (RTD), which is increasingly led at the European level, and mandated utility programmes, many governments have mechanisms to encourage private-sector investment – having already recognized that existing utilities are not the most promising proponents of renewable energy, they have proceeded to support independent investments. The Danish and British programmes were examined as case studies in Volume I, Chapter 7, showing among other things the importance of such private-sector supports. As noted in that chapter, these are not the only countries to offer such encouragement.

The situation in six European countries is summarized in Table 1.5. The structure and detail of the support varies widely but most of these systems, in one way or another, result in a premium price being offered to generators of renewable electricity.

The realized price paid for such renewable energy generation for the four EU countries with the largest energy consumption is shown in Table 1.6. The payments differ for different sources, and between the different countries. The higher tariffs for biogas and solar electricity in Germany and Italy reflect the fact that such payments are being used explicitly to promote the less developed renewables. Payments in France are notably low, in some cases less than half those in Germany and Italy; the British NFFO (Non-Fossil Fuel Obligation) system (in its third round) generally yielded intermediate prices.

Currently, most of these systems are hybrids. Except in the UK, they exist to promote renewable generation by independent companies, inputting to systems where the great majority of generation is by companies that focus

Table 1.5 Support systems for renewable energy in six European countries

	Germany	Italy	Denmark	Netherlands	UK	Spain
Reserved market	No	Yes	No[a]	No	Yes[a]	No[a]
Price basis	Reference to average sale price: 90% for PV; 75% for wind and biomass; 65% for mini-hydro	Reference to 300 MW CCGT plant	Reference to domestic price; tax relief for wind and biomass	No standard price. Low-voltage buyback at 90% sale price. Premium on buy-back price for certain distributors	Competitive bidding within fixed bands. 'Strike price' offered to each selected bid	Reference to sale. Price set to include externalities. Additional incentives for renewables since 1995
Duration	Implicit	8-year contracts	Subject to agreement	Implicit	15-year contracts (since 3rd round)	Contracts at least 5 years
Bidding	No	No	No	No	Yes	No
Selection procedure	No	Yes	No[b]	No[b]	Yes	No
Other means	Subsidies from various ministries and *Länder*. Fiscal incentives	Half connection cost for certain projects	Capital subsidies (abolished for wind energy)	Subsidy from distributors (2% tax). Some specific finance schemes	Some supports through European (ALTENER) programme	Subsidies from public funds (at 30% capital cost). Other financing being put in place

[a] Quantified objectives are set in Spain, UK and Denmark.
[b] No authorization required for up to 50 MW in Denmark, 5 MW in the Netherlands.
Source: Translated from Table 3.3 in D. Finon, *Production Décentralisée d'Electricité à Partir des Energies Renouvelables et de la Cogénération: Etude Comparative des Conditions d'Achat en Europe*, Vol. 1: Rapport de Synthèse (Grenoble: Institut d'Economie et de Politique de l'Energie, July 1995).

on large-scale centralized power generation and are integrated with, or have a special relationship with, monopoly distributors. It remains to be seen how such supports will fare if and as countries move more towards liberalization of electricity supply.

Table 1.6 Comparison of prices paid per unit for renewable energy generation in 1993–4[a]

Energy source	Germany		Italy		UK (NFFO–3)[b]		France	
	(pfennig)	(ecucents)	(lire)	(ecucents)	(pence)	(ecucents)	(cents)	(ecucents)
Mini–hydro	12.5	6.8	117	6.7	4.3–4.9	5.7–6.6	30	4.8
Wind	17.3	9.4	150	8.6	4.0–4.8	5.4–6.5	29	4.7
Biogas	14.4	7.8	222	12.7	4.9–5.2	6.6–7.1	35	5.6
PV	17.3	9.4	222	12.7	–	–	25	4.0

[a] 1993 home currencies and 1995 ecus.
[b] NFFO = Non-Fossil Fuel Obligation (see Volume I, Chapter 7). In January 1997 the UK announced prices for the fourth NFFO round, which revealed further price reductions: mini-hydro 4.25 p/kWh;large windfarms 3.53 p/kWh; and biogas 5.51 p/kWh (all capacity weighted average prices).
Source: Translated from Table 3.1 in D. Finon, op. cit. (see source to Table 1.5).

The Liberalization Directive contains explicit provisions allowing Member States to favour renewable energy sources, and other indigenous primary sources.[11] Despite this, policies that require utilities to subsidize renewables are bound to come under increasing pressure as utilities face increasing competition. Already the German utilities have challenged the legality of the supports for wind energy on the grounds that they represent a distorting subsidy – one for which, moreover, they have to pay. Thus one issue that will have to be faced in the coming years is the compatibility of the present supports with liberalized electricity markets. This is considered further in Chapter 4, on wind energy, which has become the prime focus of these disputes and issues.

At present, liberalization has not gone far enough, and the renewable electricity contributions are too small, for the competitive impacts of renewable energy payments to be an issue in most cases. There are still monopoly generators or franchised distributors upon whom the obligation to pay re-

---

[11] 'A Member State may require the system operator [transmission and/or distribution], when dispatching generating installations, to give priority to ... renewable energy sources or waste or producing combined heat and power' (*Directive Concerning Common Rules for the Internal Market in Electricity*, Articles 8 and 11). A Member State may give priority to indigenous primary fuel sources not exceeding 15% of overall electricity supply.

newable-energy tariffs can be placed, without fear of customers deserting to generators that do not face such obligations; and the sums involved are tiny in comparison with overall system costs. Ultimately, however, if liberalization proceeds to its logical conclusion and as renewable energy capacity grows, different or more harmonized systems of support may need to be introduced, as considered in Chapter 8.

In principle, this issue is not unique to renewables. Supports for nuclear power and coal have come under increasing scrutiny already, and similar questions seem likely to arise for any efforts to deflect investments away from the wholesale dash-for-gas that otherwise seems likely to follow fuller liberalization of European electricity and gas systems. But for renewables particularly, the structure of supports will need to be developed in recognition of the particular nature of renewable sources in European electricity systems. This relates not just to their characteristics as nascent and generally environmentally desirable sources, but also to a wider set of technical characteristics that we consider in the next chapter.

Chapter 2

The Role of Variable and Dispersed Sources in Electricity Systems

Small island and other isolated systems, which supply probably 2–3 per cent of Europe's population, offer a particularly promising initial market for renewables because most depend for supplies upon diesel or other oil-fired generation which is more expensive than bulk generation on grids. The variability of renewables is a drawback but enough experience has been gained to design integrated control systems to optimize fuel savings, including combinations of renewables, storage capacity and load management. Such applications are limited but could form a useful niche market for some renewables. The need now is for a compendium of options and experience and associated system modelling tools, and greater field experience and case studies, as a basis for a targeted strategy for renewables-based remote supplies, drawing on the Union's Structural Funds.

For grid-connected applications concerns about the variability of primary renewable electricity sources have been greatly exaggerated. The positive seasonal correlation between electricity demand and hydro, wind and wave energy in northern Europe is an advantage, and the lack of short-term correlation at times of peak demand is not a significant drawback; system studies confirm that contributions up to 5–10% of total electricity supply are just as valuable as those from 'firm' baseload sources and may be more so. The larger dispersal implicit in larger capacities reduces the variations, and the generation value declines only slowly with increasing penetration; it is improbable that significant generation penalties will arise from the variability of such sources. The situation is more complex for PV, but the positive short-term correlation with electricity demand in specific end-use applications, notably some buildings, can increase its value relative to conventional sources.

Over one-third of the costs of many European power systems are associated with distribution networks. Most primary renewable technologies are small enough to be connected at the level of distribution systems, potentially saving on transmission, conversion and distribution costs and losses higher in the system. The added value of such 'distributed generation' will vary widely according to site and application. In general, the added value may be in the range 5–20% as compared with centralized generation, but it could be more for applications with storage, or in remote (e.g. mountainous) regions, where transmission lines have to be buried, or for PV at the point of end-use. Current structural and technological trends point towards the emergence of 'distributed resource' systems, in which generation may occur at many different points in the network at many different scales, yielding much greater sensitivity to network costs and consequent matching of local generation with demand. The emergence of such distributed resource systems would thus assist renewables, particularly photovoltaics, which could be used to strengthen weak points in networks and which may be matched to loads at the point of end-use.

The adequacy of monitoring and control systems, planning capabilities, and institutional structures for distributed resource systems still needs to be resolved. However, liberalization will increase the access for independent renewable electricity generation, and could and should be framed to increase the opportunities for taking advantage of local matches between renewable energy and demand in rural and remote parts of networks, and other potential niche markets.

2.1 Introduction

The renewable sources of 'primary electricity' – those such as wind, solar, hydro, wave and tidal energy that produce electricity directly from mechanical or photoelectric conversion – differ from most conventional power sources in several important ways. Their output is 'variable': it follows the fluctuations of the natural cycles. They are usually (except for large tidal and wave energy schemes) available on much smaller scales; as such they can be installed on relatively short timescales and would usually connect to distribution networks rather than feed directly into the high-voltage transmission system. Finally, they are cheap to operate once constructed; the main cost lies in construction.

This chapter examines the potential role of such renewable sources in power systems, with particular emphasis upon the first two characteristics. It also explores in more depth possibilities for the evolution of 'distributed resource' electricity systems, as providing a broader context for the assimilation of dispersed renewable sources.

2.2 Islands and other isolated supplies

Islands and other isolated supply systems form a natural initial focus of attention for renewable energy investments, for two reasons.[1] First, many such locations have relatively good renewable resources, from one or more of the following natural features: strong winds; shorelines for wave energy; high rainfall and relief for small hydro power installations and/or strong sunlight. Indeed, the southernmost areas of the EU, most suitable for solar energy, are all islands. Second, the cost of conventional energy supply tends to be much higher than for grid-connected supplies because they cannot benefit from economies of scale.[2] Frequently also they rely on oil-based power – diesel or small thermal oil stations – because of the greater ease with which these fuels can be transported and used for fuelling stations of smaller capacities. The greater use of renewables in such systems therefore benefits from competing against higher-cost sources and contributes to reducing long-term oil dependence. Even where island electricity grids are, or can reasonably be, connected by submarine cables to mainland supplies (as in Crete), this nevertheless involves significant costs and some operational losses.

Volume I notes that about 4% of the Union's population lives on islands (other than the UK and Ireland). Even allowing for some islands which are

---

[1] The EU's Electricity Liberalization Directive defines a 'small isolated system' as one with 'consumption less than 2500 GWh in the year 1996, where less than 5% of annual consumption is obtained through interconnection with other systems' (Directive 96/92/EC, 19 December 1996, OJ no. L 27/20, 30 January 1997). This is over 0.1% of total EU electricity supply and equates to perhaps 500 MW capacity – a definition which encompasses almost all island and remote supplies except for those connected to mainland grids.

[2] Although technical developments have lessened economies of scale, as noted below, by reducing the cost of smaller generating stations, this applies most strongly to gas which – because of its own economies of scale in delivery – is rarely available on islands.

Box 2.1 Renewables for isolated households and small villages

Until the 1990s, probably the most widely held image of renewable energy was of isolated houses, or perhaps villages, complete with their own windmill and solar panel: a sign of self-sufficiency, isolated from the national grid. In some cases, this is now a credible option, but it is only likely to be economic for locations some distance from a grid, or where the density of electricity demand is very low. Figure 2.1 shows how the costs of grid connection vary according to the number of consumers per kilometre of line in a developing country, and compares this with the cost of PV for supplying basic lighting and audiovisual needs (shaded area).

Figure 2.1 Variation of connection costs with consumer density of population

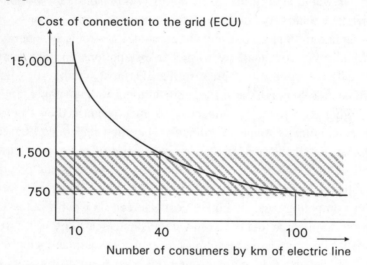

Source: Ademe, *Electricity from Sunlight, the Photovoltaic Solution* (Valbonne Sophia Antipolis, France: October 1995).

This is particularly relevant in developing countries, where grids are less developed, electricity consumption per household tends to be lower, and solar resources are good. If each consumer has low electricity consumption – just for lighting – the cost of meeting these needs by PV may be less than that of grid connection if there are fewer than 100 consumers per kilometre. At higher (but still relatively modest) consumption, including for example TV and radios, the figure might be 40 consumers per km.

Other renewables are less well suited for supply at the scale of individual households; power from wind turbines is cheapest for sizes of 100 kW or more, for example, sufficient for several hundred households. This – together with the fact that grids already connect most towns and villages in Europe – complicates any assessment of the potential for isolated supplies, especially for European conditions, which must also take account of seasonal variations and other aspects of off-grid supply. Specific data for PV are given in Chapter 5.

well integrated into mainland systems, this still represents an important niche market for nascent renewable energy industries. One study states that five million people – about 1.5% of the EU-15 population – live on islands with their own local electricity supply.[3] Given the lower-than-average per-capita consumption in such applications, fully isolated systems probably account for under 1% of total Union electricity supplies, but the relative contribution to the quality of life on islands and other isolated situations – where energy supply is recognized as one of the constraining factors – could be far greater than this suggests.[4] The potential role of the various renewable sources thus needs to be more fully developed. This is also an obvious candidate for integration with use of the Union's regional Structural Funds.

Various technical issues need to be addressed, among them that of systems integration. For small systems, the variability of some renewable sources may be a substantial drawback. The output from just a few wind turbines clustered on one site, for example, may generate a substantial fraction of the total power. The output may be very variable, sometimes fluctuating widely within a few minutes. The rest of the system, which might well amount to a few diesel stations, has to alter its output rapidly to follow the changes, with individual units repeatedly shut down and restarted, which can waste a great deal of fuel and increase maintenance requirements. Careful control strategies – often involving a small amount of storage – can help to ensure system stability and to optimize fuel savings.[5]

---

[3] Per D. Andersen, Peter H. Jenssen, and Brian J. Jensen, 'Electricity from renewables: ALTENER project AL/71/93/DK', in *Renewable Energy Entering the 21st Century*, Proc. ALTENER Conference, Sitges, Barcelona, November 1996 (Brussels: EC-DGXVII).

[4] S. Gouvello et al. argue that 'when local managers of conventional rural electrification have the possibility of balancing conventional grid and decentralized solutions, there emerges a much larger-than-expected potential for local energy projects', citing institutional reasons why utilities avoid classifying sites as non-connectable. Govello et al., *Institutional Bottlenecks Limiting the Diffusion of Renewable Energies* (Paris: CIRED, 1996), p. 26.

[5] The most complex control issues arise on small wind-diesel systems. For an overview and references see R. Hunter et al. (eds), *Wind-Diesel Systems – A Guide to the Technology and its Implementation*, prepared under the auspices of the International Energy Agency, Cambridge University Press, 1994; also a brief summary in M. Grubb and N. Meyer, 'Wind energy: resources, systems and regional strategies', in T.B. Johansson et al. (eds), *Renewable Energy* (Washington DC: Island Press, 1993), pp. 176–8.

Nevertheless, the high cost of alternative supplies may well justify such projects. Extensive research has explored many different options, and applications of wind-diesel systems in particular are growing rapidly.[6] Various forms of short-term storage can greatly improve the characteristics; in such isolated applications it may make sense to use a combination of primary renewable sources for supplying electricity and heat in combination with heat storage and/or load management. Much relevant knowledge and experience can be found in thousands of scattered conference papers; but the principal RTD need now probably relates to greater field experience and tools to aid case-specific design. In the early 1990s, the industrial experience grew, with the emergence of companies specializing in advanced power and control technologies coming together with generators to market 'intelligent' power systems for decentralized electricity supplies.[7]

In the developing world, the scope for such supplies is huge: they represent applications which are in themselves important for economic development as well as environmental protection. In Europe, the market for isolated supplies is relatively small but significant in terms of providing a base for some renewable industries (notably photovoltaic solar cells, PV). However, the great bulk of consumption of fossil fuels for power generation is in systems integrated with the ubiquitous national grids, where different issues arise.

2.3 The value of variable sources in integrated power systems: generation impacts

In integrated systems, the cost of conventional power generation is much lower than for isolated supplies, and therefore a harder target against which renewables must compete. On the other hand, there are fewer problems in integrating variable sources in such systems. Concern is frequently expressed about the variable nature of renewable sources, which depend directly on the vagaries of winds, sun, river flow, tides or waves for their

---

[6] Hunter et al. (eds), op cit., list more than thirty wind-diesel systems and analyse several in detailed case studies.

[7] For example, the joint venture between SMA, a German power electronics and system design company, and Powercorp, an Australian company specializing in isolated power supplies.

output. In fact, in contrast to the use of such sources in isolated systems, the penalties associated with the impact of variable sources in large integrated power systems turn out to be relatively minor.

If variable sources account for only a small part of the installed capacity – a 'small system penetration' – fluctuations are lost among the variations in the demand for electricity; in fact it can be shown mathematically (as well as demonstrated in simulations) that the impacts on scheduling of other plant and the need for reserves are negligible if the variable source supplies just a few per cent of demand.[8] The 'operating' penalty in fact remains small – a few per cent of fuel savings at most – for contributions of probably well over 10% of system energy. This is partly because of the inherent diversity and inertia of large systems. They have a greater capacity for absorbing variations, and most have hydroelectric units or gas turbines that can respond quickly as conditions change on the network. Significant renewable capacities would involve many different units, spread out over many sites, which tend to smooth out the variations: relative swings in the power are reduced, predictability is increased, and the overall distribution becomes much more favourable with far fewer occasions of near-zero or peak output.

A related concern is about short-term stability impacts. This is an issue both of bulk system response and of networks (considered below). A major European integration study has explored this and other aspects of integration in relation to wind energy, and concludes that in most cases system stability considerations do not pose undue problems for wind energy even though this is the least predictable source.[9] The main issues therefore concern generation availability, and transmission and distribution costs and benefits.

---

[8] This discussion is drawn from a broader study of the issues involved in integrating renewable electricity sources into power systems, including the possible role of 'limited-energy plants' and the issues concerning smaller systems, in M.J. Grubb, 'The integration of renewable electricity sources in power systems', *Energy Policy* (June 1991). A technical analysis is given in M.J. Grubb, 'The value of variable sources on power systems', *IEE Proceedings C* (London: Institution of Electrical Engineers, February 1991).

[9] F. Walker, 'Integration of new energy in utility networks', *Proc. Euroforum on New Energies*, UNESCO, Paris, 5–6 July 1993. However the point continues to be widely misunderstood, even in the Energy Directorate's most recent publication, which repeats, without any analysis or justification, concerns about the systemic implications of their 'renewables-intensive' scenario (*Energy In Europe*, Special Issue (Spring 1996), p. 35).

There still could be occasions without power from variable sources, so that some 'backup' is still required if the capacity of variable sources rises much above the level of general statistical fluctuations in peak demand and the availability of conventional power stations.[10] Yet this backup need not be provided by storage. Technologies such as gas turbines can respond quickly to meet peaks and are relatively cheap to build. Connections with other power systems can also prove a source of peak reserve – one of the major reasons for existing interconnections in Europe. Hydro power stations and biomass electricity systems may also provide opportunities for cheap backup, because although it may be costly to increase the total energy available, the incremental cost of adding generating capacity alone may be low.[11] The key questions therefore do not concern backup options considered in isolation, but rather the overall capital, operating and environmental costs and savings, taking into account the need for adequate security of system operation. Extensive use of variable sources increases the value of storage, and storage in turn may increase the value of renewable sources, but storage is in no sense a central element.

These issues have been examined extensively elsewhere, and examples relating to particular sources are given in the respective chapters.[12] One important issue is the *correlation* between the variations in renewable output and those (of electricity demand and the availability of other plant) in the rest of the system. In most of Europe, and particularly in northern Europe, there is a substantial positive seasonal correlation between electricity demand, small hydro, and wind and wave energy, and a negative seasonal correlation with solar energy. The short-term relationship is more variable and

---

[10] In practice, the capacity of a variable source typically has to rise to around 5% of the installed capacity of thermal stations before the statistics of its variation become significantly different from those of conventional power stations – which can also fail. In practice very few plants are really 'firm'. Thermal plants can fail, losing several hundred megawatts of generation at short notice, which is one of the reasons why systems have to carry spare capacity anyway. Variable sources can therefore serve to displace some conventional capacity without loss of reliability.

[11] Technically, these are known as 'energy limited' plants.

[12] Walker, op. cit. (note 9); a detailed comparison of wind energy modelling studies is given in M.J. Grubb, 'The Economic Value of Wind Energy at Higher Power System Penetrations: An Analysis of Models, Sensitivities and Assumptions', *Wind Engineering,* Vol. 12, No. 1, 1988.

can depend upon the site and application, as further discussed in the following chapters of this report. The British example is of interest because it illustrates a neutral case in which there is limited short-term correlation between electricity demand and renewable energy sources. Primarily on the basis of modelling studies of wind and tidal energy in the British electricity system, it has been concluded that:[13]

- The electricity from variable sources is usually as valuable as that from conventional stations that generate an equivalent amount of electricity, in terms of both fuel and capacity savings, when the variable source forms a modest fraction (up to 5–10%) of the total generating capacity.
- It seems highly unlikely that the use of variable sources will ever be seriously inhibited by bulk system limitations. In many cases, and taking into account the likely evolution of power systems in response to rising levels of variable sources, contributions of perhaps 20% of the demand could be obtained from one type of variable source with only a modest (less than 10%) reduction in the value of the electricity as compared with 'firm' baseload plant. Modelling studies suggest that an integrated British system could in theory – and neglecting siting constraints – accept contributions of 40–50% of electricity from wind energy before the operating penalties become necessarily intolerable, even without additional storage or possible power exchanges with other systems. By using combinations of different variable sources, hydro, storage, and/or trade, there seems no technical reason why large systems should not derive well over half of their power from variable sources.

It is uncertain how far the quantitative conclusions apply if the only significant renewable source is PV. Since the output from installations several hundreds of kilometres apart could still be closely correlated, large capacities would involve correspondingly large power swings, with significant power produced for perhaps only one-third of the time. The economic limits on PV would depend strongly on the diurnal and seasonal characteristics of demand on the system, and storage with a capacity sufficient to replace much of the lost supply, at least across the evening peak, might make a large

---

[13] Ibid.

difference. Even without such storage, however, considerable amounts of fossil fuel could be displaced by PV if systems developed to accommodate it.

With the greater availability of natural gas, and growing interest in gasification of biomass fuels for power generation, in the long term power systems could develop a substantial resource of plant with low emissions and low capital cost, but moderate to high fuel costs. These are exactly the characteristics required for low-cost 'backup' of variable power sources. To put it another way, if the system has much plant which is cheap to build but expensive to run, variable sources are very valuable just as fuel savers. Their short-term value in reducing coal and oil bills could in principle be transformed into a long-term role of helping further to limit emissions and extend gas and/or limited biomass applications.

So, although the variability of renewable electricity sources is often cited as a major obstacle, in reality it seems likely to be one of the least significant technical constraints when such sources are integrated into large power systems. However, since this capacity to assimilate variable sources arises from the properties of the system overall, there is a danger that competition in electricity supply will be introduced in a way that does not reflect the real value of variable sources, particularly those relating to capacity payments (as has partially occurred in the UK). The fine print of regulatory change therefore needs to be scrutinized carefully even with respect to the value accorded to variable generation. This applies even more importantly to issues of transmission and distribution, on which the rest of the chapter now focuses.

2.4 Network costs

The transmission and distribution side of the business is just as relevant as generation, for two reasons. First, some of the greatest concentrations of renewables are far from demand centres; for these, the costs and losses associated with long-distance transmission must be considered. Second, more locally available renewables would, unlike the traditional power sources, feed into systems at the level of distribution networks rather than directly to bulk power transmission. The potential significance of this is hard to estimate for a particular location and even harder to generalize about. This section provides some background information relating to

Table 2.1 Breakdown of ENEL electricity cost

Stage of formation	Cost (%)		Percentage of total
	Capital	Operation	
Generation	21.7	78.3	54.3
Transmission	60.0	40.0	1.7
Distribution			
At high voltage (60–50 kV)	33.5	66.5	6.5
At medium voltage (2–60 kV)	43.6	56.4	13.2
At low voltage (220 V)	41.9	58.1	24.3

Source: ENEL.

overall distribution costs; the following sections explore the possible implications for the economics of localized generation.

One indication of the importance of (low- and medium-voltage) distribution costs is given by investment expenditure by major utilities. A breakdown of investment costs for the Italian utility ENEL (Table 2.1) shows that 37.5% of expenditure was on distribution, which is comparable with total expenditure on the operation of thermal power generation. Investment in transmission and high-voltage distribution totalled around 8%. This indicates that little over 50% of the cost is associated with generation. Transmission and distribution at medium voltage, and distribution at low voltage, account for 21% and 24% of the delivered cost respectively; but only a small fraction of this cost (1.7% of the total) was high-voltage transmission. Such data do not appear greatly atypical; for smaller industrial consumers in England and Wales, distribution charges account for about 28% of charges while transmission accounts for about 6%.[14]

Thus at the very large-scale end of the business, bulk electricity transmission cost is a small fraction of total costs. For many applications, transmission costs seem set to decline further. In particular, technologies for high-voltage DC transmission have steadily improved. Thus the Norwegian transmission company Statnett has embarked upon ambitious plans to export power across the North Sea to Germany and to Denmark, landing it still at highly competitive prices, as discussed in Chapter 7.

---

[14] *The Supply Price Proposals* (London: Office of Electricity Regulation, 1993).

Table 2.2 Connection voltages for sources of different power output (UK standard)

Power level (MW)	Voltage level (kV)
100–700	132
10–50	33
0.5–20	11
0.3–1.0	0.415–0.220

Source: Integration of renewable energy sources in electrical power systems, *Watt Committee Report on Renewable Energy Sources* (Watt Committee, 1990).

At the level of distribution the costs are far higher and the outlook is, perhaps surprisingly, more problematic, particularly for rural areas. In France, in which rural demand has risen (despite the decline in population) to account for over 15% of total demand, expenditure on rural distribution reinforcement alone accounts for about 1 billion ecus, and the per-unit cost of that supply is estimated to be more than twice that of supplying urban centres.[15]

Table 2.2 shows the typical voltage connection levels for sources of different sizes, illustrating that installations of a few megawatts such as wind farms would generally be connected at 11 kV or sometimes 33 kV, whereas individual turbines and other smaller systems could sometimes be connected at even lower distribution levels. To the extent that such sources meet demands on the local network rather than going through bulk transmission, this will increase their value.

An alternative indication of the importance of 'network costs' is the gulf between bulk generation costs and delivered prices, apparent from Table 1.3. The delivered price to domestic users is generally more than twice the nominal cost of generation. Clearly, a central issue for renewable power sources is whether they are competing against the price of major thermal stations – or the price of electricity delivered to dispersed users.

---

[15] Christophe de Gouvello and Marcello Poppe, *Maîtrise de la Demande d'Electricité et Surcoûts de la Desserte Electrique Rurale*, CNRS No.940 (Paris: CIRED, 1994).

2.5 The value of dispersed generation

As indicated, many renewables are small enough to be connected to distribution networks rather than to the main transmission system. How much is this worth?

The answer depends upon the connection voltage, and upon source location and output relative to local demand and other power sources. If there is no local network, or demand is negligible compared with the source output, the power will have to be stepped up directly to bulk transmission, with the associated capital costs and conversion losses clearly assigned to that source. Similarly, if the source is located on a distribution network on which local generation already exceeds the demand, power has to be transformed up to the transmission system and there are additional losses as compared with centralized generation. On modern systems, the latter situation, particularly, would be unusual. The power from most renewable sources would go first to meeting demand on local networks, thus avoiding – as compared with centralized generation – the operating costs and losses associated with transmission, conversion and distribution down to that level; it just reduces demand on the relevant substation.

Early evidence based on a handful of studies indicates that in ideal circumstances the value of PV placed at key points in the distribution system can be more than twice the value of centralized power production. An analysis made by the US utility Pacific Gas & Electric on its Kerman substation has shown this. This means that PV generation could cost twice as much as conventional central station generation, and still be cost-effective in carefully selected applications. However, the unique nature of the local conditions favouring this remarkable result must be emphasized. Such conditions are difficult to reproduce for other systems or even in other locations within PG&E's own network.[16]

---

[16] See Chapter 5. The PG&E load in that region is largely determined by air conditioning, and hence is correlated with solar radiation intensity and is almost synchronized with PV output. It was this that enabled capacity credits, both at local T&D and system level, that would be inconceivable for the vast majority of European utilities. For the latter, the peak period is in winter, during the hours when PV generation is minimal or indeed nil. Moreover, the meteorological conditions of PG&E's chosen site enable use of the plant – and, accordingly, energy credits – that are unarguably superior (+30%) to those usually obtainable at the best southern European sites.

The data above suggest that more typically, for sources feeding into distribution at medium voltages, distribution savings might imply a premium perhaps of the order of 5–20% as compared with bulk generation; however, the uncertainty and variability of any such credits must be emphasized. Advantages for very small sources at low distribution voltages – almost at the point of end-use – could be considerably higher. Actual savings from any particular installation depend upon the state of the local network and whether upgrading of the transformers would otherwise be considered. If the source output is firm or correlates well with local demand, or is combined with local storage, there may be capital network savings associated with permanently reduced demand at higher levels of the network; otherwise, in general, there will not be. If at times the local generation exceeds the local network demand, the operational 'savings' for those periods will be transformed into the costs associated with stepping up to bulk power transmission. The location of the source relative to demand centres on the local network will affect losses within that network; and so will the location of the distribution network within the overall system. Notably, demand areas at the end of long feeder lines – for example, mountainous regions or other sparsely populated areas – may involve losses far greater than the system average, and the value of local generation will be correspondingly enhanced.[17]

One trend that may increase the value of such distributed generation is the rising environmental pressures on distribution itself. New distribution lines in Europe attract increasing opposition, primarily because of their visual impact but also because of fears (however slim the evidence) about the health impact of electromagnetic fields. In many countries, indeed, it is almost impossible to gain new 'wayleaves' for new distribution lines, and even upgrading is difficult in sensitive areas. Increasingly, utilities are being driven to place distribution lines in such areas underground, which can increase costs by a factor of ten. In these circumstances, if distributed generation can reduce the need for new or reinforced distribution lines, its value may be considerably increased – though the same sensitivities could increase planning obstacles, particularly for wind power.

---

[17] All these factors are amplified by the fact that line and transformer losses increase in proportion to the distance and as the square of current; the impact of marginal changes is thus greater than the average costs and losses suggest.

The two central questions facing renewables in this respect are first, technically, can electricity systems develop to absorb and make full use of the potential benefits of dispersed generation? And second, can regulatory and planning systems evolve to enable the generators to receive the benefit, and if so, how? This takes us into a broader debate about the structure of future electricity systems.

2.6 'Distributed resource' systems

The discussion above illustrates that the value of power at a small scale may differ markedly from that of bulk power generation, and that this value may vary greatly according to the location on the system and timing of output. As economies of scale no longer dominate planning, other factors, including line losses, transmission upgrade costs, reliability etc., assume greater importance. The changing face of electricity supply is thus not merely confined to issues of scale and organization leading to greater competition. To match the theoretical engineering discussion above, a concept of 'distributed resource' systems is emerging.

In this concept, central station production would be augmented by rapidly developing 'distributed resource' technologies. These comprise not just renewable energies such as photovoltaics and wind, but also fuel cells, small gas-fired or diesel-engine generators, battery and other small-scale storage, and customer efficiency/load management devices located in the distribution system. If distributed technologies are cost-effective on a broad scale, this may change significantly the way future systems will supply and deliver power and energy to their customers. This in itself may be a technical driving force behind potential changes in utility structure and operation. Power will be generated with a wider range of energy resources and technologies, some of which could even be mobile for relocation as the power needs of customers change or until more permanent assets are economically justified.

Economic advantages

Location of generation near or at the point of use derives some additional economic value by reducing the transmission and distribution costs required to deliver the electricity. Modular technologies also can be installed in incremental capacities that meet the demand at the time much more closely. In a sense, the system may be approaching a 'just-in-time' philosophy to adding new capacity.

This incremental construction approach improves the effectiveness of capital by putting plant quickly into service with little excess capacity. It also permits the rapid adoption of technological changes. Moreover, the functions of many distributed supply technologies will be more than generation alone. They can increase the reliability and power quality of local supply, improve the load factor on transmission and distribution (T&D) assets and displace additional T&D investments. These attributes give the distributed resource (DR) technologies substantial added economic benefit.

Over and above this, much of the value of distributed technologies is in their modularity and in the managerial flexibility they allow. The previous chapter noted the lowered and uncertain prospects for electricity demand. Traditional utility planning involves decisions about very expensive projects which last 30 or more years, based on cost projections over that period that almost invariably prove to be wrong. DR technologies open options that were not available before. For example, expansion for a rapidly growing area could be met initially by adopting DR technologies, enabling decisions on bigger investments to be deferred and re-evaluated as trends and costs become clearer. This managerial flexibility of DR technologies may be valuable, though its quantification is complex (see Chapter 8).

Technology and engineering aspects

As noted, the distributed resource system would emerge as a mix of central station and distributed generation together with demand-side management technologies, connected to the grid at strategic points or located at or near loads on either side of the meter (Figure 2.2). Candidate technologies include primarily PV, small motor-generator sets and energy storage, such as batteries and superconducting magnetic energy storage. Other candidates are smaller installations of technologies such as wind turbines, solar thermal systems and fuel cells. Demand-side management options include hundreds

Box 2.2 ENEL's preliminary experience in distributed applications of PV

Because of its modularity, PV generation is well suited to installation in the medium- and low-voltage distribution network, making available distributed economic gains in addition to other benefits. (e.g. the value of the energy produced, generation capacity credits, environmental improvements, etc.). Distributed benefits can be maximized if it is possible to identify special locations (or 'niches') where the reduction of power-flow peak achieved with PV generation increases economic gains.

To study this, ENEL conducted a distribution network analysis to a sample of feeders in western Sicily. The study revealed the existence of lines with both structural (length) and functional characteristics that are particularly critical and that, under proper circumstances, can be relieved by PV generation.

These included lines 100–150 km long, with a maximum distance from the substation bus-bars of about 40–50 km; lines characterized by losses (in power) in excess of 3% (up to 8–9%), and voltage drops greater than 5% (up to 9–11%); and lines characterized by 500–1,000 forced interruptions per year (for single and multiphase failures) and by 40–80 tripping operations per year for maximum current.

Studies showed that on some such lines, incorporating PV generating 1000 kW in summer and 800 kW in winter could reduce line losses by 200–300 MWh per year, which translates into an economic benefit of 14–21,000 ecus, equivalent to an overall present value (calculated over 25 years at 5%) of 183–305,000 ecus. In particular, one line is notable for the possibility of PV generation reducing losses by over 470 MWh/yr with an annual benefit of 33,000 ecus, equivalent to an overall present value of about 477,000 ecus.

These advantages must, naturally, be added to the (more substantial) savings in generating costs from conventional plants: preliminary valuations suggest this would be about 77,000 ecus/yr, with overall present value of 1.1 million ecus.

A number of these lines have loads that peak outside the insolation hours, or an evening peak the same size as the morning one. This is the case for those lines serving mainly domestic users. Without local storage, the effect of PV generation is obtained only during the daytime peak (when it could more than halve the peak voltage drop).

For lines which have a typical industrial load diagram (two daytime peaks), the benefit of reducing the voltage drop is limited to the extent that PV generation displaces the load peak into the evening. In doing so, the reduction in voltage drop is only in the order of 7% to 6.1%. Taking into consideration the current growth rate of consumption in the Sciacca zone (4–5% per year), the installation of 1000 kW of PV generation could enable postponement of the replacement or installation of a transformer for two years. The total saving should be in the order of 44,000 ecus.

of building technologies, appliances and other consumer technologies, and industrial process technologies used for load management. Unlike conventional generation technologies such as large coal and nuclear plants, the emerging technologies tend to be small and modular, achieving their cost efficiencies through economies of product, or mass production. For

Figure 2.2 Schematic comparison of centralized and distributed electricity systems

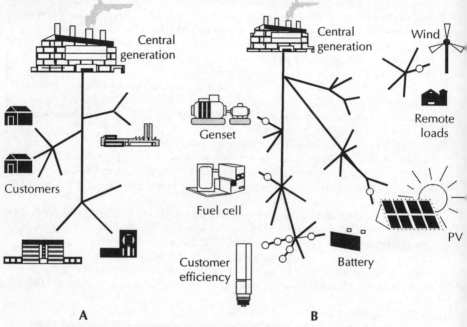

A B

Source: J. Ianucci and J. Eyer, 'Thoughts on the distributed utility and implications for India', in P. V. Ramana and Keith Kozloff (eds), *Renewable Energy Development in India: Analysis of US Policy Experience*, New Delhi: TERI; Washington: WRI 1995.

example, instead of making on-site generator sets, wind machines, solar PV, or even high-efficiency compact fluorescent lamps bigger to reduce costs, the factories make more of them. This concept requires a fundamental change of mindset for the utility industry, which is familiar with capturing economies in field construction rather than in factory manufacturing.

Many technical issues associated with the physical operation of distributed generating and storage technologies remain to be clarified. How are they to be operated? Would 'islanding' operation be permissible in certain segments of the network? If not, what interface devices are envisaged to avoid it? According to what logic will the protection devices operate? How should the networks be redesigned? Should maintenance procedures be modified? If so, what new skills will be needed by the relevant staff?

Accordingly, there is little experience of systems with a medium or high penetration of distributed technologies. Below a certain level, system operation could continue as usual. As penetration increases, however, the unprepared utility may find itself unable to utilize, control, monitor or dispatch the new resources effectively, or perhaps at all, without major changes. If inadequately managed, the new generation could affect system voltage and protection adversely, and DR technologies could also undermine the role and interests of traditional utilities. Customers could use many of the new modular technologies for self-generation, leaving the grid altogether, or using the rest of the system for backup, which would result in a cost burden to other customers.

Institutional and regulatory aspects

These new technologies, for both electricity companies and customers, also affect the way the electricity is considered as a product. In the past, any price differentiation was based on broad market segmentation, e.g. into residential, commercial, industrial and so on. In short, the consumer treated electricity as a commodity. Likewise the 'centralized' utility produced, delivered and sold its product as a commodity.

With the emergence of distributed options, customers and system operators alike may seek energy pricing differentials that take into account time of use, reliability, power quality (to protect new consumer products, such as microprocessors) and other factors – fully 'cost-reflective' pricing, varying according to location, time and other factors. The market could move towards a more discerning customer who wants unbundled services, and the new technologies can allow the system to deliver them. Taking this trend to the limit and considering that the customer buys energy for light, space conditioning, torque and the like, the ultimate product might be these unbundled services.

Grid-connected and isolated on-site generation and energy efficiency could also be among these services. Although third parties, not utilities, might actually perform the services, the system operator is in a good position to be the market aggregator and financier because of its current close relationship with its customers and the knowledge it has about them. In some countries, competitive bidding and regulatory incentives for customer energy efficiency are already taking utilities down the road to this new way of business.

To all of this is to be added the considerable value attributed to environmental benefits and to the substantial estimated gains in terms of service quality improvement. However, an important constraining relationship should also be recognized between the distributed resource system, distribution issues and the broader issues of integration with generation on large systems discussed earlier. Unless there is a significant correlation between load and generation, reliability and overall system control can only be maintained by resorting periodically to the centralized production system's stabilizing capacity, to the detriment of distributed benefits.

Given the conditions under which distributed technologies yield greatest value, efforts must clearly be addressed to identifying the most favourable sites. This is a difficult task, and currently available generation planning tools take limited account of the geographical distribution of electricity demand. The true value of distributed generation can (at present) be determined only through detailed and cumbersome 'manual' analyses such as those used by PG&E to study the Kerman substation example above. There are currently no off-the-shelf planning tools to assist in the design of the generation system while taking into account the multivariate needs of the distribution system or the stability of the transmission system. Current planning models, complex as they are, do not provide good – if indeed any – handling of naturally varying output from sources such as wind and PV, and incorporating distributed storage is even more complex.

Indeed, the concepts of distributed resource systems themselves raise new regulatory issues about operation, ownership, control and transparency of costs that will need to be resolved, and that are discussed further in Chapter 8 of this volume.

2.7 Conclusions

The value accorded to renewable electricity sources – and hence their investment prospects – depends upon a wide array of regulatory and technical issues governing the nature and operation of European electricity systems. These span the traditional issues of investment criteria and potential tax or

Box 2.3 Key characteristics of the 20th- and 21st-century electricity systems in Europe

Key characteristics of the European electricity system	
20th century	21st century
Few generating technologies	Many generating technologies
Inflexible generating technologies	Flexible technologies
Simple electricity system control technologies	Complex control systems
Increasing demand	Static, falling or increasing demand
Increasing economies of scale	Static economies of scale
	Increasing economies of scope and system
Consumer pays – government responsibility	Justification to investors and customers
Limited risk	Increased risk
Captive customers	Customer choice
Geographical imperative of siting	Siting for efficiency
Long lead times for new power plant	Short lead times
Power plants on stand-by	Short start-up times
	Environmental concerns
Bus-bar costing methodology	Cost-reflective pricing

Source: C. Mitchell (University of Sussex, September 1996).

other incentives, the trend towards liberalization, and the minutiae of regulations which determine how the system evaluates and pays for independent generation and takes account of the potential benefits associated with dispersed renewable electricity generation.

Two broad and closely related themes emerge from the analysis of trends in European electricity systems: the institutional changes associated with liberalization; and the technical changes associated with the rise of smaller-scale generating, control, storage and load management technologies that open the possibility of 'distributed resource' systems. Together, these are leading to an important array of changes that may distinguish electricity systems in the next century, as summarized in Box 2.3.

Technical trends are leading to systems with a much greater choice of generating technologies, many of them relatively small with short lead times and start-up times and hence quite flexible; many of these could benefit from economies of scope and system rather than scale. Competitive trends are leading to a situation in which generators face increased risk because they can no longer automatically pass costs through to customers, and investments have to be justified to financiers, shareholders and customers; this leads to a preference for plants with short lead times and greater environmental sensitivity, given rising environmental concerns and greater customer choice.

This chapter has shown that large capacities of variable renewable electricity sources can be incorporated even in traditional electricity systems. The addition of much more sophisticated control systems, the greater diversity of technologies, and the use of cost-reflective pricing should further increase the ability of power systems to absorb variable sources in the most cost-effective ways. Greater flexibility in siting should benefit renewables, most of which are by their nature dispersed, and the relatively small scale of most renewables fits well with several of the other characteristics set out in Box 2.3. Thus, most renewable electricity technologies fit better with the structure and methodologies that are beginning to emerge in European power systems. Especially if legislation is formed with a careful eye to ensure that renewable generation is accorded its full value in the system, trends in European electricity would appear to favour renewables in many respects.

Chapter 3

Hydro Power

Hydro power is the most familiar and best developed of the renewable electricity sources. Large hydro schemes (above 10MW) already supply 13% of European electricity, and an economic potential exists to extend this to over 18% in the EU-15, primarily from resources in southern Europe, Austria and Sweden; the remaining potential in Norway is comparable. However, although many environmental impacts can be mitigated by good design, public opposition and the higher cost of capital in more liberalized systems are likely to limit practical expansion to a much lower percentage.

Small hydro stations have generally less environmental impact, but tend to be more expensive per unit capacity and provide less storage to the system. Currently, small hydro plants supply 1% of EU electricity, and this could plausibly expand to 3–5%, with refurbishment, retrofitting and the development of new sites. Electricity from many small hydro schemes may have relatively high systemic value as distributed generation, and hence may benefit if liberalization develops with cost-reflective pricing. Local planning procedures are an important determinant of the prospects for such exploitation.

The potential for hydro power exploitation in central and eastern Europe, and in Iceland, amounts to several hundred TWh/yr. Liberalization of European electricity could improve the prospects for imports, and for EU financing to develop such resources, but would inhibit larger developments especially because of the higher cost of capital. Such investments could be drawn from funds for assisting the transition in central and east European countries, but hydro has not figured previously in such expenditures, probably because of the long timescales involved. With adequate funding, imports to the EU-15 totalling up to an additional 5% of EU electricity consumption might be feasible, generating sustainable hard-currency revenues for these regions. Further afield, there are also huge surplus resources in west Africa and in the Caucasus, but development of either faces major obstacles given the high investment and political risk involved.

Globally, the biggest remaining potential for hydro power is now in the developing world, and European industry dominates the global market. Continuing development and refurbishment of hydro power in Europe, along with promotion of cleaner energy investments world-wide, could contribute to maintaining this position.

3.1 Introduction

Hydro power is the most established and familiar renewable electricity source. World-wide it accounts for 18% of all power generation, and in Europe almost 14%; at present it dwarfs all other renewable electricity contributions combined. Because hydro power is a well-established and familiar technology, the scope for further expansion in Europe is relatively limited, especially for large schemes, and the policy and industrial issues are more familiar than those for other renewable energy sources.

3.2 The hydro resource and environmental considerations

The hydro resource can be quantified with greater meaning than for many renewables, though some uncertainty arises as to what is 'feasible' economically and environmentally. Table 3.1 shows the estimated 'economically feasible' potential for the EU, European Economic Area (EEA) and central east European countries compared with current hydro generation in the most recent year reported. The data are divided between large- and small-scale hydro schemes – a distinction usually taken by international convention to be delimited at 10 MW but in some EU countries demarcated at 5 MW or 1 MW – because the resource and policy issues vary greatly between 'big' and 'small' schemes.

With the addition of the hydro-rich systems of Sweden, Finland and Austria, large hydro power contributes 13% of EU-15 electricity at present. Exploiting the economically feasible resource would increase this to over 18%. This remaining potential is located primarily in southern Europe (Greece, Italy and Portugal), and in Austria and Sweden. Additional resources immediately on the EU's border are dominated by Norway, with almost as much additional economic potential as in the whole of the EU-15.

Large hydro schemes tend to be economically attractive, and offer useful regulatory and storage capacity for power systems, but a range of social and environmental impacts arise from the damming of the river and creation of a large reservoir, including population displacement; inundation of plants and ecosystems; changes in water quality, sometimes involving health concerns; impacts on fish arising from these changes, the altered flow patterns and the physical obstacle; and sedimentation which may both damage the reservoir quality (and availability) and contribute to changes in erosion and fertility patterns downstream.[1] There is no doubt that understanding of these problems has improved greatly and many negative impacts can be greatly reduced in modern hydro power station design, construction and operation;[2] but they remain important concerns that have become the main constraints upon exploiting the remaining resource in much of Europe.

Thus the potential for large hydro schemes in western Europe is largely exploited, and this is reflected in the fact that the annual generation is for most EU countries quite close to the estimated economically feasible potential; the difference may be ascribed to environmental and a variety of other constraints. There is some scope for expansion in southern European countries, but in Sweden and Austria it is unlikely that much more can be exploited; in Sweden, laws now prevent exploitation of the four main remaining undammed rivers on environmental grounds.

---

[1] For detailed discussion of these impacts, see Jose R. Moreira and Alan D. Poole, 'Hydropower and its constraints', in T.B. Johansson et al., *Renewable Energy* (Washington DC: Island Press, 1993). Some case studies of environmental impact are summarized in Geoffrey P. Sims, 'Hydroelectric energy', *Energy Policy*, Vol. 19, No. 8 (October 1991), reprinted in Tim Jackson, *Renewable Energy: Prospects for Implementation* (UK: SEI/ Butterworth-Heinemann, 1993).

[2] 'The frequent lack of competence in handling the social and environmental aspects of hydroelectricity, especially in large-scale projects, has contributed to an exaggerated pessimism in many quarters regarding the intrinsic negative impacts of hydropower. Such a situation can be blamed in part on the traditional ethos of the electric utilities, which are dominated by engineering and financial priorities ... [especially] ... in developing countries and in the eastern bloc countries where most hydroelectric potential exists ...', Moreira and Poole, op. cit.

Table 3.1 Hydro resources and contributions[a] in European countries

	Large-scale hydro		Small-scale hydro[c]		
	Economically feasible resource[b] (TWh)	Current generation (TWh)	Exploitable resource (TWh)	Current generation (TWh)	Percentage of total electricity generation
EU-15	**430.2**	**292.5**	**> 97.5**	**> 24.58**	**13.9**
Austria	53.7	34.0	3.1	5.10	70.5
Benelux	0.6	0.5	n.a.[d]	0.44	0.3
Denmark	0.1	0.0	n.a.	n.a.	0.016
Greenland	14.0	0.2	n.a.	n.a.	40.6
Finland	13.6	13.5	1.5	1.02	23.2
France	72.0[e]	69.2	12.0	7.65	14
Germany	20.0	18.3	n.a.	n.a.	3.4
Greece	16.0	2.9	2.0	0.12	8.2
Ireland	1.2	0.8	n.a.	0.03	5
Italy	65.0	42.3	65.0	6.87	19.5
Portugal	19.8	8.9	6.5	0.23	32.3
Spain	54.0[e]	25.1	7.0	2.84	15.7
Sweden	95.0	73.3	n.a.	n.a.	52
UK	5.2	3.5	0.4	0.28	0.1
European Economic Area	**281**	**147.7**	**58.5**	**5.65**	**85.3**
Iceland	44.0	4.5	45.0	0.15	95
Norway	200.0	111.7	8.0	2.99	99.6
Switzerland	37.0	31.5	5.5	2.51	61
Central-east Europe	**240.5**	**> 65**	**> 3.5**	**> 3.5**	**7.6**
Albania	17.0	5.2	n.a.	n.a.	90
Bulgaria	10.5	1.9	0.8	0.45	5.9
Croatia	n.a.	3.7	n.a.	0.08	37
Czech and Slovak Republics[e]	10.8	2.5	0.2	0.53	3
Hungary	3.5	0.2	0.3	0.03	0.5
Poland	n.a.	1.2	n.a.	0.58	1
Romania	17.0	10.9	0.4	0.93	17
Slovenia	7.0	3.0	n.a.	0.19	31
Turkey	124.6	36.4	1.8	0.44	38
Former Yugoslavia[e]	50.1	n.a.	n.a.	n.a.	28.8
Former USSR	**409.45**	**> 167.0**	**> 357.0**	**> 0.12**	**n.a.**
Russia[e]	167[e]	167.0	357.0	0.12	20.6
Baltic states	6.15	n.a.	n.a.	n.a.	n.a.
Western CIS	50.6	n.a.	n.a.	n.a.	n.a.
Southern CIS	185.7	n.a.	n.a.	n.a.	n.a.

[a] Contributions of current generation for 1994.
[b] 'Economically feasible hydropower capacity' based on local conditions.
[c] Data derived assuming that 1 MW produces 0.0034 TWh/yr.
[d] n.a., Data not available
[e] Indicates 'technically feasible hydropower capacity', which is total potential irrespective of economic or environmental considerations.
Source: Hydro data: *International Water Power and Dam Construction, Handbook 1995* (London: Reed Publishing, 1995). Percentage contribution derived from IEA data on electricity production.

Small hydro plants (less than 10 MW capacity) contribute about 1% of European electricity supplies, from an installed capacity of around 4000 MW. The main attractions are the lower environmental impacts, and the fact that small hydro schemes may offer supplies for isolated, island or mountain locations. However, they offer less storage and regulatory capacity. Many small schemes are 'run of river' without dams, and the power output may fluctuate considerably over periods of days to weeks. Small hydro also tends to be more expensive (per unit output) to construct.[3]

In addition to new schemes, there is said to be significant scope for refurbishing old plant to increase output, and installing turbines in existing, non-generating dams;[4] no data were found on the potential.

The main untapped resources lie beyond the EU's boundaries. To the north is Norway – already exporting hydro power through the Scandinavian system and with internal debates over the desirability of developing more – and Iceland, where the constraint is the absence of any market.

To the east, the central-east European countries do not have great potential for increasing large hydro generation though the resource is significant, particularly in the Balkans. The biggest potential arises even further east –

---

[3] 'The capital costs of any new hydro station will vary depending upon design, location and rating ... but as a good rule of thumb, an installed price of 1800 ecus/kW for a plant above, say, 50 MW would be a reasonable figure to take. But below ratings of 10 MW, the unit capital cost rises quite dramatically. This rise has to do with the physics ... and not any limitation in equipment manufacturer or designer capabilities ... below 500 kW prices may be as high as 4,000 to 6,000 ecus/kW'. From working group report on small hydro, in *Action Plan for Renewable Energy Sources in Europe* (1994), p. 7.
[4] For example, Sims, 'Hydroelectric energy', op. cit., and Moreira and Poole, 'Hydropower and its constraints' op. cit., which discusses costs and motivations for refurbishment.

most notably in Turkey and Georgia – and there are massive potential resources in Russia and some of the former Soviet South.

To the south, there is some hydro potential in the Atlas Mountains, but the biggest unexploited African hydro resources lie in central-west Africa, notably proposed 20,000 MW developments at Grand Inga in Zaire, and other possible projects on the Congo river.[5] Routes for transmitting such power to Europe have been proposed via Tunisia, Morocco or Turkey, or some combination thereof. Such large-scale interconnections are considered further in Chapter 7.

The pattern for small hydro is less marked, and the data and definitions are less clear and consistent. The TERES study[6] and the European Small Hydro Association (ESHA) estimate an exploitable resource of around 80 TWh. The economic potential for small hydro thus may amount to around 4% of European electricity supply, of which only about one-quarter has been exploited to date. This is a significant resource and industry. The ALTENER programme sets a goal for doubling the contribution by 2005, and the ESHA calls for this to be doubled again by 2020. The quality of data is poorer for east European countries, but there appears to be considerable potential for small hydro in the Balkans, and presumably more widely in mountainous regions.

Broadly, therefore, the policy issues divide into two main areas: the potential for further development of small hydro schemes, particularly within the European Union; and the scope for developing large hydro resources around the periphery.

3.3 Expanding the small hydro contribution

Hydro power, large and small, is a mature technology sponsored by a mature industry, but incremental technical improvements are still occurring and more are possible. For large hydro these comprise various ways of lowering the cost of dam construction, and of minimizing environmental impacts. For small hydro, the development of mass production of components has helped

---

[5] 'Remote renewable energy resources: long-distance high-voltage interconnection', *IEEE Power Engineering Review*, Vol. 12, No. 6 (June 1992).
[6] The European Renewable Energy Study, EC DG-XVII (Brussels: 1994).

substantially to narrow the 'economy of scale' cost gap. A number of novel
ideas for generating power from small hydro sites have been advanced. An
Energy Technology Support Unit (ETSU) evaluation states that these do not
offer significant advantages over improved conventional designs,[7] and the
European Action Plan does not suggest the need for a coordinated European
RTD effort.[8] However, EC assessments in 1996 did conclude that 'there are
still opportunities for RTD in all features of the technology', and successive
rounds of the UK's NFFO system have seen declining bid prices for small
hydro, along with rising contracted capacity.[9]

The constraints upon greater exploitation of small hydro power in the
EU, and the policy issues involved, have been well rehearsed. Some constraints are genuine physical and economic ones: many sites may not be
near demand, and the costs and/or environmental impact of transmitting
modest amounts of electricity to the grid may be prohibitive. In these conditions small hydro nevertheless must be a natural candidate for remote supply if there are nearby villages not connected to the grid. This may well be
both economically and environmentally preferable to grid extension over
mountainous terrain, though the quality of supply may be lower. In view of
the continuing emphasis upon grid extension in European power systems
(Chapter 2), far more attention may still need to be paid to the possibility of
smaller hydro systems for remote supply as an alternative.

For small hydro plants within reach of grids, conversely, the terms of
access for independent generators are obviously important. The UK NFFO
has demonstrated the continuing scope for small hydro developments by
independent players; 15 new schemes were awarded contracts in 1994
under the first Scottish renewables order.[10] Switzerland has embarked upon
a two-million-ecu programme to expand and upgrade its small hydro
resource.[11] In these and other programmes, designs emphasize continuity of
river flow, unobtrusiveness, management of fish stocks etc.; the Swiss

---

[7] *DTI Small Hydro Programme Review* (Harwell: ETSU, September 1995).
[8] Working group report on small hydro, op. cit. (Note 3).
[9] Mark Allington, ETSU, personal communications.
[10] For an overview see E. Henfield, 'Small scale hydropower in Scotland', *CADDET Newsletter* (Paris: IEA, 1996).
[11] 'Launch of the DIANE small hydro project', *CADDET Newsletter* (Paris: IEA, July 1995).

programme also highlighted the potential for refurbishing old schemes, and the 'astonishingly high' potential available in drinking-water supply systems from Swiss mountain lakes.

There are various other barriers to small hydro power. The industry lays great emphasis upon the problems caused by lengthy and bureaucratic procedures for authorization in relationship to land planning and ownership, environmental impact assessment, fishing rights and water abstraction permission (and sometimes charges). Conditions vary widely across the Union and there is limited scope for Commission intervention, except perhaps in the form of guidelines concerning maximum acceptable delays, and principles governing elements of Environmental Impact Assessments and water abstraction. The industry notes that the lack of European-level standards still hampers the market: 'There remains the need both to harmonize the available standards at the European level, and even for new basic standards and codes of practice to be developed.'[12] These and some of the regulatory principles are being considered under the ALTENER process.

The greatest economic determinants are the costs of finance and the value that can be obtained for the electricity generated. As illustrated in Chapter 1, small hydro plants benefit from the buy-back arrangements for independent power producers in most countries, but at somewhat lower rates than other renewables, presumably because of the desire to focus on stimulating development of newer technologies. However, small hydro is often likely to be a relatively valuable power source. With annual load factors typically of 65–75% reaching 90% during peak winter months, hydro can usefully contribute not only a bulk system 'capacity credit', but also local system benefits in the form of embedded generation. Given its predominant occurrence in mountainous regions or plateaus that are also relatively expensive to supply by grid – and that are generally not near gas grids for local gas generation either – small hydro could be a major beneficiary of the liberalization of European electricity systems, *provided* that this is done in such a way as to reflect the benefits of distributed generation (Chapter 2). Combinations of wind and hydro power could also be attractive, given the potential (albeit limited) storage and regulating capacity of small hydro schemes.

---

[12]Working group report on small hydro, op. cit. (Note 3), p. 6.

As for other renewables, however, the financing of such small and relatively capital-intensive projects can be costly, and the price and availability of long-term loans is thus another determining issue.

3.4 Development of large hydro resources outside the EU

Outside the EU, similar issues for small hydro stations remain important but the scope extends also to large hydro. This raises two broad policy issues. One is the adequacy of infrastructure for delivering such power to markets. The planned sub-sea cables from southern Norway to Germany and the Netherlands (see Chapter 7) will give central Europe access to some of the Norwegian hydro resource, and earn Norway foreign exchange and potentially 'carbon-credits'.

Exploiting more of the resources in central and eastern Europe would generate useful amounts of clean power for these regions, and strengthened east–west power lines might open the possibility of developing resources further afield for the European market, in the process earning these countries much-needed foreign exchange. Apparently little attention has been given to these resources in the loans of the European Investment Bank and the European Bank for Reconstruction and Development, or in the PHARE and TACIS programmes, probably because they require projecting and financing beyond the next few years.

The progressive liberalization of European power systems is also relevant to the exploitation of large hydro resources outside the EU, but in a very different way from small hydro. For developed and financially strong countries such as Norway, liberalization opens the door to power exports much more freely than before. In terms of tapping resources elsewhere – such as in central and eastern Europe – it raises the possibility of large European companies financing hydro power developments in other countries, to transport the power across European grids for their own consumption.

As discussed in Volume I, Chapter 5, the central and east European countries are mostly not short of power but they are short of finance, and economic activity more generally. Given the scale of the unexploited hydro resources, particularly in Albania and the former Yugoslavia where evaluations and projects may have been rather limited by political circumstances, the scope for such investment as part of reconstruction efforts – aided by the PHARE programme – deserves further exploration. Developments further east – power from the Balkans, Turkey, western Russia, and possibly even further out to Georgia and the southern republics – could in principle proceed on the same basis, through the TACIS programme, though the uncertain political situation is an important impediment. Similar remarks apply to the potential development of African resources for exports to Europe.

Another possibility for remote hydro power development is in Iceland, which has for years been promoting a scheme to develop its hydro and geothermal resources for export via sub-sea cables to Scotland. This is discussed in the broader context of possible North Sea developments in Chapter 7. Overall, the scale of hydro power imports to the EU is hard to project, but a long-term figure of around 100 TWh/yr, or up to 4% of EU-15 supplies appears plausible.

Liberalization, however, is by no means a boon to hydro power. The economics of large hydro power (and its relative, tidal power) – which particularly but not exclusively involve large schemes with long lead times – highlight probably more than in the case of any other renewable technology the tremendous importance of capital financing conditions. Almost all the cost is incurred up-front, with long lead times for large hydro, after which the plants have very long lives of generating power with very low operating costs. The terms of financing, and the question of credit for long-term operation, echo like a mantra throughout the literature on hydro (and tidal) power. The issues are generic and are discussed more fully in Volume IV of this series, but the prospects for further development of large hydro power for imports will be uniquely sensitive to the availability of long-term and low-interest finance.

Finally, the potential for further development of hydro power in the developing world is very large. In generation, the power would go to meeting the explosive growth of demand in these countries for the foreseeable future. European industry already dominates export markets both for large- and small-scale hydro equipment. Further developments in the home market – along with support for clean energy investments world-wide – would presumably help to maintain this position.

Chapter 4

Wind Energy

Wind power has developed very rapidly since the mid-1970s as a technology and industry. The major developments occurred in Denmark and California in response to government policies which supported renewable technologies and resource surveys and encouraged private-sector power generation. In response to government incentives in many different European countries, capacity in Europe has expanded by 25–30%/yr during the 1990s. By late 1996, wind energy supplied more than 4% of Danish electricity, and capacity across Europe exceeded 4000 MW. Generating costs vary according to financing and siting conditions; costs in European conditions in 1995 spanned 5–9 ∈/kWh and have continued to fall. The scope for further technical improvements and market expansion suggests long-run generating costs in the range 2.5–5.5 ∈/kWh depending primarily on site conditions and interest rates.

The diversity of large wind energy contributions would help to smooth the output and make it relatively predictable. Regional contributions exceeding 10% of electricity supply from wind energy can be accommodated with little loss in economic value; indeed for contributions up to 5–10%, the seasonal correlation between wind energy and electricity demand may enhance its value compared with equivalent output from conventional sources. For smaller contributions, in appropriate locations, this may be further enhanced by the distributed benefits associated with supplying local demands. Much larger regional contributions would incur increasing distributional and operating penalties, though contributions exceeding 30–40% of demand are technically possible without having to discard any electricity.

Wind energy resources are strong in many of the more remote areas of the EU, including mountain regions, Greek islands, and many coastal regions especially along the Atlantic and North Sea. Indeed almost all the Celtic regions of western Europe, from Galicia in the Iberian peninsula, to Ireland

and Scotland, combine high wind resources with relatively poor economic development and often moderate to high electricity prices. Wind energy in these regions and much of Greece, potentially aided by the use of structural funds, represents a stable long-term basis for the large-scale development of European wind energy generation and industries. Better data are needed concerning wind energy in central/eastern Europe.

Siting guidelines can overcome potential impacts of noise, etc. The main constraint is visual impact, especially in hilly terrain. Estimates of credible resources are highly subjective because of their dependence upon site availability, but extrapolation from existing experience and detailed surveys suggests that a capacity of perhaps 50–100,000 MW, supplying 5–10% of European electricity, is likely from sites on the European mainland and shallow coastal waters. Most of this could be constructed in the period 2000–2020, making a substantial contribution to energy diversity and environmental goals on these timescales. Accessing larger resources around the European periphery in the long term probably depends upon development of submarine connections through the North Sea, perhaps connecting with Scottish wind and Norwegian wind and hydro resources. Offshore wind energy adds further to the available resources, and further development, including trial deep-water wind farms, should be supported.

Acceptance of wind energy on a large scale hinges upon perceptions and land-use planning policies. Perceptions in Europe are mostly more positive than in the US despite some concerted opposition, particularly in the UK and Germany. The key to long-term expansion is integration of wind energy within established local land-use planning procedures, so that wind energy proceeds at a measured pace with the support of local populations and institutions.

Wind energy has prospered on the basis of government-sponsored incentives providing low-cost capital and premium rates for power generated. The latter mechanisms have repeatedly been challenged by utilities in Continental Europe, and wind energy, as the most visible and rapidly growing of the renewables, has borne the brunt of such attacks. These supports will come under increasing pressure as liberalization proceeds and reliance on direct premium payments must decline as the industry matures.

European wind energy industries dominate the global scene, and the continuing tragedy of wind energy in the US is Europe's opportunity. However, fragmentation of European industries and the lack of involvement by Europe's major engineering companies limit the financial strength of European industries. The protection of national wind industries by many European governments has outlived its usefulness, and needs to be replaced by open competition in procurement. A strategy for wind energy in Europe thus has five main components:

- *continuing RTD in association with manufacturers, including offshore developments;*
- *competition in procurement within national support mechanisms;*
- *establishment of credible targets, consistent with international environmental commitments and backed by matching support mechanisms including European structural funds;*
- *establishment of more appropriate premium-payment regimes through the transition of liberalization;*
- *integration of wind energy into other national and regional policies, particularly local land-use development plans and international aid and export finance.*

On such a basis, wind energy is likely to become a major European power resource over the coming 20 years, and an important European export industry.

4.1 Introduction

Wind energy is the most commercially developed of all of the 'modernized' renewable electricity technologies. In addition to traditional uses for isolated applications, already by 1990 there were more than 20,000 electricity-producing wind turbines connected to mainland electricity grid systems, mostly in the United States but with growing numbers in Europe. Wind energy in 1990 supplied about 2% of electricity consumption in California and Denmark. Subsequent developments have been very diverse – the Danish proportion rose to 4% by 1996, while that in California actually declined.

World capacity in mid-1996 exceeded 5,000 MW.[1] Installations in Europe have soared past those in the United States, on the way to a capacity likely to exceed 5,000 MW by 2000, and European manufacturers dominate the international market. What can be learned from the wind energy experience, and what are the future prospects and strategies which may be developed?

This chapter examines the potential evolution of the wind energy industry in Europe, and the main policy issues which arise, in somewhat greater depth than for other technologies considered in this study, given wind's pre-eminent place among the newly developed renewables in Europe. The chapter first outlines key characteristics of wind turbines as a power source, and then sketches the developments that have brought wind energy to its current position; Sections 4.5 and 4.6 examine more closely the costs of wind energy and the potential for improvement. Section 4.7 reviews wind energy resources in Europe, and Section 4.8 assesses the economic place of wind energy within Europe's power systems. Then the two key dimensions of investment decisions – planning applications and tariff structures – are considered in more detail. Finally the chapter offers a more strategic view of the wind power industry in Europe and its prospects internationally.

4.2 The characteristics of wind energy[2]

Almost all commercially available wind turbines for bulk power generation consist of two or three blades, mounted on a horizontal hub at the top of a tower which also supports the gearing and power conversion equipment. Commercial machines range in height up to about 40 m (hub height above ground level) with capacities up to 750 kW, but there have been experimental machines several times this size, and commercially available machines increase in capacity almost every year; several companies successfully marketed machines up to 1.5 MW in 1996.

The primary attractions of wind energy are its environmental advantages and the relatively small unit size. The main obstacles are the relative youth

---

[1] *Windpower Monthly*, October 1996.
[2] For a full review of wind energy technology, characteristics and development, the reader is referred to Paul Gipe's excellent book, *Wind Energy Comes of Age* (New York: John Wiley, 1995).

of the technology in its modern form, its capital-intensive nature, and the visual and occasionally other local impacts of wind turbines; there may be consequent difficulties in siting large numbers. The variable nature of the winds can pose difficulties for applications on remote sites or weak grids. These issues are discussed below.

The environmental attractions of wind energy are the same as for most renewable sources. It produces no solid or liquid wastes, and no gaseous emissions. It requires no external fuels and so avoids the environmental problems of fuel extraction and transport, as well as the economic dangers of fuel supply interruption and price hikes. Siting is not dependent on the availability of cooling water, and land can be shared with other applications such as farming; typically, only about 1% of the area of European wind-farms is actually occupied by the base and access routes.

Potential environmental drawbacks include electromagnetic interference, noise and visual impact. To avoid interference with TV and other transmissions, wind turbines have to be sited away from main transmitters, and they can interfere if sited on a line-of-sight of microwave communications (e.g. radar and field telephones). Reception interference can also occur, especially in areas with already poor reception, but a local relay station to amplify the signal, or cable connections to affected houses, can be installed at a small fraction of the windfarm cost. Electromagnetic interference is therefore not considered a significant obstacle, and no sites in Denmark have been refused on these grounds.[3]

The noise from wind turbines places a limit on how close to houses they can be sited. There is considerable variation according to machine design and location.[4] Problems with unusually noisy machines have led to their withdrawal from the market, but noise still constrains the available sites; regulations in Denmark restrict noise levels at dwellings to 45 dB, a level which is 'greater than in a bedroom at night, but less than in a house during

---

[3] M. Grubb and N. Meyer, 'Wind energy: resources, systems and regional strategies', in T.B. Johansson et al. (eds), *Renewable Energy* (Washington DC: Island Press, 1993).

[4] Difficulties in measuring and interpreting noise data are discussed by A. Robson, 'Environmental aspects of large scale wind power systems in the UK', *IEE Proceedings A,* Vol. 127, No. 5 (London: IEE, 1980); issues of *Windpower Monthly* report occasional public complaints against particular installations.

the day'.[5] With current technology this requires a distance of about 300 m for single machines, or up to 500 m for a windfarm.[6] Studies have not generally revealed significant problems with bird deaths but this has emerged as an issue in some locations, particularly in relation to the local presence of birds of prey and/or migratory routes.[7]

The most common objection raised against wind energy is that of visual impact. For wind energy to make a substantial contribution, thousands of large wind turbines would be required, and so they would have to be visually acceptable (see Section 4.10).

The relatively small unit capacity of wind turbines compared with most power plants opens the technology to relatively small investors and enables very rapid installation, usually within a few months of signing contracts. It allows production-line methods to be used, with fixed-price contracts for standardized products, and the primary risk is borne by the manufacturer rather than the purchaser.

Furthermore, when problems do occur, it is usually possible to replace faulty components relatively quickly. Modern wind turbines therefore tend to be technically more available than conventional plant; commercial windfarms installed since the late 1980s have typically registered more than 95% annual availability over several years of operation; indeed, anything short of this is now considered newsworthy in the trade press.

Winds fluctuate on all timescales. As outlined in Chapter 2, the value of such sources in large power systems does not depend much on having reliable output at times of peak demand, nor do variable sources impose significant operational penalties, because there is a statistical contribution to reliability and because plants such as pumped storage or gas turbines are used to provide protection against load fluctuations. But the broader relationship

---

[5] European Wind Energy Association, *Wind Energy in Europe – Time for Action* (Oxford: EWEA, October 1991).

[6] M. Grubb and N. Meyer, op. cit. (Note 3).

[7] 'Birds ruling affects 8,000 turbines', *Windpower Monthly*, January 1991; 'Bird deaths study reveals true cause for alarm', *Windpower Monthly* (May 1991). Press reports such as these led to extensive investigations, reviewed in P. Gipe, op. cit. (Note 2). These indicate that problems do exist for birds of prey in some US locations (though rarely to an 'ecologically significant' degree). Studies in Europe, including those led by conservation organizations, have generally allayed fears (see various issues of *Windpower Monthly*).

between wind output and electricity demand can be important, and is discussed below.

4.3 The international development of wind energy to the mid-1990s

There have been several periods of research interest in 'modern' wind energy since the beginning of the century, but the main developments did not start until the oil price shock of 1973. The subsequent evolution of wind energy for grid supplies has occurred in four distinct phases. The period 1976–81 saw a range of government programmes, notably in the United States, Sweden, Germany and Canada, aimed at developing very large turbines and understanding the underlying technology. Although much was learnt, most projects ran into substantial technical problems and high costs.

The second phase, from 1982 to 1985, was dominated by the development of a market for small- and medium-sized machines in the United States. As outlined in Volume I, Chapter 7, a favourable regulatory regime combined with generous federal and state tax incentives in the early 1980s made wind energy in some areas – particularly California – an attractive private investment, even at the high costs then prevailing. Installed capacity in California rose from about 10 MW in 1981 to over 1000 MW by 1986, when all the major incentives were withdrawn. The cumulative investment in Californian wind energy by 1986 totalled about $2000 million,[8] with the value of the energy generated put at $100 million/yr.[9] These developments were a mixed blessing, because the incentives were so high that they encouraged deployment of inadequate machines in haphazard and poorly planned windfarms. Although some companies had made major strides with their technology, the reputation of wind energy in California suffered and installation rates collapsed after 1986.

With the market base and finance of California, several companies invested heavily in wind energy technology and gained rapid experience. The results over this period were a doubling in the mean size of commercial

---

[8] In 1986 the exchange rate between the dollar and the ecu was about US$1 = 1 ecu.
[9] P. Gipe, 'Maturation of the US wind industry', *Public Utilities Fortnightly* (20 February 1986).

units to over 100 kW; major improvements in machine performance; and a rapid fall in capital costs, from US$3,100/kW in 1981 to an estimated US$1,250/kW average in 1986 (historical prices).[10]

The third phase of wind development, from late 1985, was dominated by the removal of US tax credits and the fall in oil prices. This greatly tightened the market at a time when several large companies had put substantial capital into new machines. The resulting pressure led to further cost and price cuts, this time driven by market pressure as much as technological change. In combination with the decline of the dollar (which greatly reduced revenues for non-US manufacturers in the US market) a number of companies went bankrupt and others merged.

The fourth phase, emerging in the late 1980s, has been one of resumed expansion, with an initial focus on Europe. The Danish target to install 100 MW of utility wind generation by the end of the decade was met with a rush in 1988 and 1989, and with more than this being installed by non-utility generators. Tens of megawatts were installed in the Netherlands by 1990 in pursuit of a target of 100 MW by 1992 (a target not met on time owing to various administrative and planning wrangles) and 1000 MW by 2000. In 1990 Germany declared a goal of 100 MW by 1995, a target which was doubled a few months later. In 1988 the UK announced plans for its first three windfarms, totalling 24 MW, proposals which were joined by other initiatives under the 'non-fossil quota' of the newly privatized electricity system in 1990. In 1991, the scale of activity expanded dramatically, with nearly 200 MW of wind energy contracts being admitted under the non-fossil quota (or about 85 MW DNC).[11] Along with this, the turn of the decade saw some revival of activity in the United States, and activity in India and a number of other developing countries.

The technological trend during this period up to about 1992 was not one of falling machine prices – which at the beginning of the period were driven by market pressures to commercially unsustainable levels – but of steadily

---

[10] Ibid.; *Wind Directions* (March 1987). A graphic illustration of the changes in the Californian market and technology is given by M. Grubb, 'Wind Energy', in M. Grubb and J. Walker, *Emerging Energy Technologies* (London/Aldershot: RIIA/Dartmouth, 1992).

[11] Figures cited variously as 82–85 MW of declared net capacity (DNC), which roughly reflects average output taking into account the variability of the wind. See Volume I, Chapter 7.

improving performance and increased size, yielding higher output per site and lower on-site costs. Most commercial machines in 1985 were under 100 kW; by 1990 most manufacturers had units of 300–400 kW. The poor performance of machines from the early 1980s gave way to availabilities of 95% and capacity factors (average output as a fraction of peak power) up to 30%, double the levels five years earlier (see also Table 4.4 below). From the early 1990s, with a more stable technological and industrial base in expanding European markets, prices began falling again, while unit size continued to increase. By 1995, units of 500 kW were commonly available and some manufacturers had ventured to 1 MW.

The contribution of wind energy is still small compared with total electricity supplies. On the relatively small Danish system, the 730 MW installed by late 1996 supplied over 4% of national electricity demand, but this is unique. In California, wind's contribution has not passed about 2% of supplies; elsewhere the contribution in 1996 was still under 1% of national electricity supply.[12] Yet the demand for wind energy is now large enough to support a substantial industry and to prompt investments to drive the technology further; and the growth rate, averaging globally about 25% annually since 1990, is by energy standards spectacular.

4.4 European developments in the 1990s and plans for 2000

In terms of capacity at least, wind energy was the greatest renewable energy beneficiary of the ascendancy of environmental concerns in Europe in the late 1980s. Following the early lead of Denmark (set out in Volume I, Chapter 7) and to a lesser extent the Netherlands, in the early 1990s Germany, the UK, Spain and subsequently most European countries introduced some kind of market support for renewable energy generation. Support regimes in the biggest countries have been summarized in Section 1.7.

Wind energy was well placed to take full advantage of these incentives. From a modest base in 1990, major developments in Germany, Denmark,

---

[12] Californian generation in 1990 was about 2,500 GWh, or 1.9% of the total Californian electricity generation of 132 TWh in 1989. Danish electricity consumption is about 30 TWh/yr, of which over 4% was derived from 700 MW of wind energy in 1996.

the UK and Spain led to a total installed capacity in Europe by the end of 1994 of 1723 MW, almost identical to the US installed capacity that year (Table 4.1). In 1995, while the US industry continued to languish and despite a pause in UK and Danish developments, the European capacity expanded by 30%, driven primarily by a surge in Germany and Spain, and with the Netherlands also becoming more active.[13] In 1995, however, for the first time in the 1990s, the European dominance of new investments was rivalled by explosive growth in Asian activity, particularly in India but also in China.

The development of the European wind energy market in total has shown remarkable consistency in the 1990s, but this masks considerable diversity of developments in member states. Danish installation rates slumped in 1992–3, following reductions in incentives and increased public, utility and political opposition, at about the time that the market in the UK and then in Germany took off. UK installations declined in 1994–5, following the long pause between the second and third rounds of the Non-Fossil Fuel Obligation, while rapidly expanding German installation was joined by Spanish expansion. In late 1996, the German boom slowed, but growth in the UK and Denmark has resumed, joined particularly by Greece. The diversity of European policies and situations, in terms of the timing of policy developments, has thus been one of their great strengths.

Many governments in Europe have established targets for renewable energy including wind energy. For the year 2000, together they total over 6340 MW, which could generate about 1% of EU-15 electricity supply and which represents a 60% increase on projections and targets of 4000 MW derived in the early 1990s. It remains unclear how realistic all these targets are, but several seem obviously on the way to being met or exceeded. The most ambitious country remains Denmark – one of the few in which energy responsibilities now lie unambiguously with the environment ministry. After the hiatus of 1992–3, debates over planning and fiscal systems and utility objections were slowly resolved, leading to resumed expansion. In 1996 the government established a national goal of 1500 MW by 2005, aim-

---

[13] Details of national programmes can be found in IEA, *Wind Energy Annual Reports* (Paris: International Energy Agency).

Table 4.1 Installed wind energy capacity in 1994–6 and projections to 2000 (MW)

Country/Region	Capacity at the end of 1994[a]	1995[b]	1996[b]	Target for 2000[a]
Germany	632	1136	1500	2000
Denmark	539	619	733	1000
Netherlands	162	236	277	1000
UK	170	200	269	800
Spain	73	145[c]	215[c]	800
Sweden	40	67	100	240
Greece	36	28	28	200
Italy	22	25	25	100
Portugal	9	13[d]	13[d]	60
Ireland	8	7	7	150
Finland	4	7	8	50
Other European	28	35	41	440
Europe	**1723**	**2518**	**3216**	**6340**
USA	1722	1655	1660	2800
Canada	na	21	21	na
Latin America	10	11	32	400
Americas	**1732**	**1687**	**1713**	**3200**
India	201	565	816	2900
China	29	44	56	730
Other Asian	7	17	24	187
Asia	**237**	**626**	**896**	**3817**
Australasia	na	12	14	
Total	**3,692**	**4,843**	**5,839**	**13,357**

[a] *Source*: 'Europe leads world wind league', *Wind Directions*, Vol. XV, No. 2 (January 1996).
[b] *Source*: *Windpower Monthly*, Vol. 13, No. 1 (January 1997).
[c] Including Canary Islands.
[d] Including Cape Verde Islands.

ing to treble Denmark's 1995 wind energy output over ten years to supply 10% of Danish electricity consumption.[14]

One analyst, Andrew Garrad, has cast the varying national situations in

---

[14] 'Domestic market revives with 1500 MW target', *Wind Directions* (London: EWEA/ BWEA, Vol. 15, No. 3, April 1996).

Figure 4.1 Stages of wind energy development

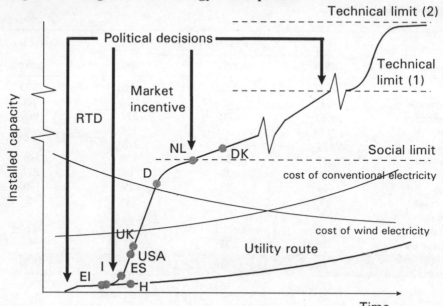

Source: A. Garrad, in *European Renewable Energy Study*, EC DG-XII (Brussels: 1992).

Europe in terms of an identifiable pattern of deployment, as illustrated in Figure 4.1. First, some capacity is installed as a part of RTD efforts, with demonstration and trial commercialization programmes run by utilities, sometimes under government mandate. The experience suggests, however, that utilities have rarely pursued wind energy with vigour or enthusiasm. Italy and Ireland, and until recently Greece, have been stuck in this phase. Rapid developments have only occurred when private generation is encouraged. Spain has started along this route, as has the UK and most spectacularly Germany. But such rapid developments inevitably spark a backlash, as the countryside in the most favourable regions sprouts wind turbines, the worst cases are publicized, and the costs of the support programme expand with the rapid increase in capacity. At this stage the rush stops, as incentives are removed or reduced and planning opposition increases. Denmark and the Netherlands have seen such retrenchment, and Garrad, along with many others, speculates that Germany is approaching such a threshold.

At some point, the industry needs to enter a more mature and stable phase of public and economic acceptability if it is to resume expansion, as appears to be happening in Denmark. Yet at present, such expansion still depends upon some kind of support or favourable tariff structure. Even if and as different European countries can make such a transition, questions remain about the economic prospects for wind energy in Europe, and strategies for its exploitation, questions to which we now turn.

4.5 Economics of wind energy

The thread of continuity between the US 'wind rush' and the developing European markets means that the critical technological lessons from the United States are embodied in current machines. The dominant technology remains the relatively simple three-bladed, fixed-speed machine, but much improved blades, controls and other components have steadily boosted performance, as indicated above. Nevertheless, there has been considerable hype, particularly in the US and probably to the detriment of the industry, as recounted graphically by the leading US industry observer, Paul Gipe:

It started in 1989 when John Schaefer published a short article in the Electric Power Research Institute's technical journal assessing the remarkable progress of California's wind industry. Schaefer concluded that at good sites wind energy cost about $0.08/kWh ... 'just about the same as that from more conventional sources'. Then, Carl Weinberg described how the cost of wind energy would fall to $0.04/kWh by 2000, one-tenth of its cost in the early 1980s. As Weinberg was head of Pacific Gas and Electric's RTD department, his projection carried clout. Not to be outdone ... in 1991 EPRI projected that the cost would drop to only $0.035 per kWh. The floodgates were open and soon a torrent of articles in the popular and trade press were discussing wind energy's new found cost effectiveness. In their eagerness, researchers and US government agencies quickly climbed aboard the bandwagon, each bent on bettering the other's projection of how cheap wind energy would become. Europeans shook their heads in disbelief. Had Americans gone berserk, or was wind energy really so inexpensive?[15]

---

[15] Paul Gipe, *Wind Energy Comes of Age*, op. cit. (Note 2), p. 226.

Gipe's conclusion is that these paper calculations, however rigorous and plausible, had a limited relationship to the conditions faced by companies in the field. In fact, it is particularly hard to generalize about the economics of wind energy. Machine costs vary according to the size and specification; costs per kilowatt have proved a poor guide as they can readily be reduced by installing a bigger generator without yielding much extra energy, so it is now more common to quote costs per square metre of swept area.[16] The costs associated with installing a machine on a site vary with the location (roughness of the terrain, distance from the grid etc.) and can add an amount anywhere in the range of 15–50% to the capital costs. Maintenance costs and land costs also vary with the site. The return depends upon the energy output, which is very sensitive to the windspeed at the site because of the cubic dependence of energy density upon windspeed. Also, the value of the energy varies considerably between systems according to other generation costs and arrangements with the utilities. Finally, because the initial capital accounts for most of the costs, the overall economics are very sensitive to financing conditions: a high discount rate will greatly increase the cost of electricity. There is no simple answer to the question 'Is wind energy economic?'

Nevertheless, existing experience gives useful indications and shows just how much the estimated costs depend on local conditions – financial and physical. Table 4.2 shows striking variations in the costs of electricity from wind turbines under 'typical' national conditions in the EU and US, according to system, financial and resource conditions. Figure 4.2 shows how costs, for typical UK investment conditions, vary with the site average windspeed – another critical factor.

In the current conditions, with cost details increasingly subject to commercial propriety, the best indications are probably bid prices in the countries where these are clear. They reflect the financial complexities, with

---

[16] 'Over-rating' of machines – installing over-sized generators to reduce apparent cost/kW – was a problem in early wind markets, and where government incentives depended on the capacity installed; this contributed to the low capacity factor (energy per unit capacity) of early machines. For machines optimally rated for a good site, the peak output is typically about 0.5 kW/m^2 of swept area, in which case the costs in $/kW are roughly twice those in $/m^2.

Table 4.2 Typical costs of wind energy in different national circumstances

	UK	USA	Denmark	Spain	Germany
Installed cost (£/kW)	900	780	800	810	1200
[ecus/kW][a]	[1086]	[941]	[965]	[977]	[1448]
Production (kWh/kW)	3000	1750	2190	2755	2200
Real interest rate (%)	9	9	5.7	n.a.	3.3
Lifetime (years)	15	20	20	n.a.	10
Energy costs (p/kWh)	4.5	5.7	3.9	6.5	7.6
[∈/kWh]	[5.4]	[6.9]	[4.7]	[7.8]	[9.2]

[a] 1995 yearly average exchange rate: ecu 1.00 = £0.83.
Source: David Milborrow, '5,000 MW of wind now operational', *Environmental Protection Bulletin*, Institute of Chemical Engineers, No. 43 (London: July 1996).

a striking range and evolution. In the second round of the UK's Non-Fossil Fuel Obligation in 1991, companies required 11 p/kWh (nearly 14 ∈/kWh in 1995 ecucents) on six-year contracts to install wind energy. In the 1994 round, with 15-year contracts and far greater confidence in the financial community, prices for larger windfarms were in the range 3.98–4.5 p/kWh (5.00–5.66 ∈/kWh), and a subsequent bid in Scotland averaged 3.9 p/kWh (4.90 ∈/kWh). In the fourth round announced in January 1997, there is 330.4 MW of large-scale wind capacity contracted in the range of 3.11–3.8 p/kWh (3.91–4.78 ∈/kWh).

Overall, two unambiguous conclusions about the economics can be drawn. First, the costs have declined steadily with expanding markets and experience, and with growing corporate and financial confidence. Second, although wind energy can already compete unaided in exceptional sites and niche applications, in general these costs are not yet low enough for it to do so in the bulk European electricity market, even in countries such as Denmark with relatively good resources and high electricity prices (though the most recent UK prices are competitive against new coal or nuclear plant). Without more direct reflections of the environmental cost of conventional sources, commercialization to date has depended upon some degree of capital credits and/or favourable tariffs for the power generated.

Figure 4.2 Variation of wind energy costs with site average wind speed

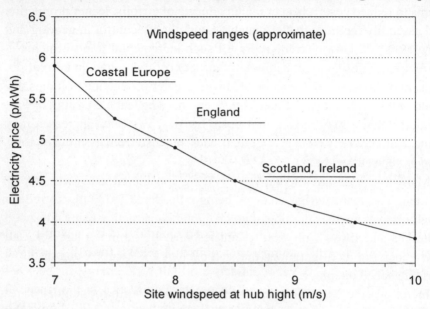

Source: David Milborrow, '5,000 MW of wind now operational', *Environmental Protection Bulletin*, Institute of Chemical Engineers, No. 43 (London: July 1996).

4.6 Potential technology developments and RTD needs

Further cost reductions may be achieved from a combination of economies of scale in production, further technical advances and developments in market conditions. Technical advances themselves can occur through a range of incremental improvements, and potentially through major discontinuous changes in design. They can seek both to increase power capture and to reduce machine costs.

Modern commercial wind turbines are 20–45 m in height with rated (maximum) power output up to 400 kW, operating at a fixed rotational speed. The Danish machines which dominate the market are designed for simplicity and strength, with three fixed blades; in most of these power is limited automatically by stalling of the blades at higher windspeeds. Some other commercial turbines have active power control by altering the pitch of the blades, and some have just two blades.

Manufacturers are continually pursuing a range of incremental improvements to reduce costs: cheaper manufacturing of airfoils, leaner production and assembly methods, better matching and quality control in gearing and generators etc. However, the room for such incremental improvements in current basic designs seems limited. Further cost reductions in current designs may be mostly those associated with greater production volumes. By analogy with other areas of mechanical engineering, each doubling of production volume may reduce unit costs by 10–15%. Such increases may come about not only from greater total sales, but also from merging of companies as discussed below.

Many minor design alternatives are being considered, and some implemented. The main design changes being considered explicitly to reduce machine costs are the adoption of much lighter construction using advanced materials and alloys with 'soft' pliant response to the major stresses, and new-generation configurations that can exclude the gearbox and/or allow variable-speed operation.

Increasing the energy capture, which also contributes to the goal of reducing energy costs, reduces relative siting costs and increases the effective resource, is of relatively greater importance by allowing more wind energy to be captured at each site. There is thus a strong case for public RTD efforts to be biased towards the goal of increasing energy capture, relative to other approaches to cost reductions. Power capture can be increased in four distinct ways: by increasing the reliability of machines; by improving the operating characteristics so that more of the incident energy is captured; by better siting including reduced interference between wind turbines; and by increasing the size and/or height of the turbines. Given current machine availability of 96–98% in commercial wind turbines there is little scope for further increasing availability.

At optimum windspeeds (typically around 8 m/s) modern wind turbines can already extract more than 80% of the theoretical maximum energy.[17] Increasing this maximum capture is difficult, though minor improvements could be achieved by further improvements in blade design. However, the

---

[17] A. Cavallo, S.M. Hock and D.R. Smith, 'Wind energy: technology and economics', in T.B. Johanson et al. (eds), *Renewable Energy*, op. cit. (Note 3).

efficiency of energy capture falls off, away from the optimum speed. Energy capture can be increased by using pitchable blades, and/or by varying the rotor speed so that it remains at an optimal ratio to the windspeed. Variable-speed operation requires a more complex electrical linkage: variable gearing; electrical transformation; or a radically different, very wide-band, generator. In the early 1990s the US Windpower company, in conjunction with US utilities, excited wide interest in the development of a variable-speed turbine with DC coupling, claiming to increase energy capture by well over 10% for little increase in machine costs.[18] But independent observers still consider it unclear whether variable-speed machines will yield cheaper energy. Other changes (e.g. better airfoils and variable-pitch operation) may also be able to extract some of the additional energy captured by variable-speed operation.[19]

Energy capture can also be increased by better siting. In flat terrain the behaviour of wind and interference between wind turbines is reasonably well understood. In hilly terrain it is much more complex, and substantial variation in output between turbines close to each other has been observed. Better understanding of siting issues is a 'common good' which does not give competitive advantage to any particular manufacturer and this strengthens the case for RTD support in this area. In the United States, EPRI estimated that better 'micrositing' may improve energy output by up to 5%.[20]

Larger wind turbines increase energy capture both by increasing the area and, at most sites, reaching the stronger winds at greater heights. The early government research programmes concentrated heavily upon large machines on the grounds that they would be essential if wind energy were to make significant contributions, but all these machines proved commercially unattractive and often unreliable because of the high stresses involved. More recent studies for the European Commission suggest that energy costs are unlikely to vary significantly for machine sizes between about 300 kW and 1 MW. On the argument that Europe needs large machines because of the limited number of sites, support totalling 7 million ecus for seven

---

[18] Ibid.

[19] A. Garrad, in *European Renewable Energy Study*, EC DG-XII (Brussels: 1992).

[20] Cavallo et al., op. cit.

Table 4.3 Large wind machines developed under EC WEGA-I and II programmes

	WEGA-I (1986–91)	WEGA-II (1991–6)
Rotor diameter (m)	55–60	50–55
Generator rating (MW)	1–2	0.75–1.2
Blade weight (ton)	6–9	2–4
Nacelle weight (ton)	80–200	35–70
Specific weight (kg/MWh per year)[a]	39–57	12–21
Specific electricity generation (kWh/yr per m²)[a]	900–1300	1200–1350
Specific electricity cost over first year of operation (ecus/kWh)[b]	not applicable	0.3–0.5

[a] At reference site, mean windspeed = 7.5m/s at 50m, no array losses.
[b] Estimated for series production of 50 units, at reference site, not including maintenance.
Source: Solar Europe, No. 3 (October 1992).

demonstration machines of around the 1 MW size (the WEGA II programme) formed the largest single item in the European Commission's renewable energy research budget in the early 1990s. Table 4.3 shows the nature and scale of developments in comparison with the EC's first large wind turbine programme. The machines constructed under WEGA-II were slightly smaller, with a variety of blade and generator and control configurations, but the weight of both blades and nacelle has in general been more than halved; in combination with increased power capture, the specific weight has been reduced by a factor of as much as three. These are stated to be demonstration machines, and some are being taken forward into commercial production. Unlike government programmes for large machines in the 1980s, these more recent efforts have built upon and extended commercial experience, and involved commercial wind energy companies.

Whereas the design and manufacture of wind turbines has become a highly specialized activity, there are elements that could benefit from more generic research and development. Power electronics can have a central role to play in enabling variable-speed operation, but also more widely for improving the power characteristics of output from wind turbines. Developments in control engineering – particularly neural networks for adaptive

Table 4.4 US wind energy cost trends and projections (in US$ or US¢) [in 1995 ecus or ecucents]

Years	1981–5	1986–91	1992–5	1996–2000	Post–2000
Installed cost ($/m²)[a]					
[ecus/m²]	650	460	400	400	350
	[617]	[419]	[365]	[305]	[268]
Availability	0.60	0.90	0.90	0.95	0.95
Operation and					
maintenance (cents/kWh)	2.5	1.5	1.1	0.7	0.5
	[2.4]	[1.4]	[1.0]	[0.5]	[0.4]
Annual energy, kWh/m² per year					
Good site, 350 W/m²[b]	350	500	630	750	750
Excellent site, 500 W/m²[c]	500	750	1025	1100	1100
Energy cost, low discount rate (cents/kWh)					
Good site	18.3	9.5	6.7	5.4	4.7
	[17.4]	[8.7]	[6.1]	[4.1]	[3.6]
Excellent site	13.6	6.9	4.6	4.0	3.4
	[12.9]	[6.3]	[4.2]	[3.1]	[2.6]
Energy costs, high discount rate (cents/kWh)[d]					
Good site	27.4	14.0	9.8	8.1	7.0
	[26.0]	[12.8]	[8.9]	[6.2]	[5.4]
Excellent site	20.0	9.9	6.6	5.8	5.0
	[19.0	[9.0]	[6.0]	[4.4]	[3/8]

[a] Lifetime 25 years; land rent 0.3 c/kWh; annual insurance = capital x 0.5%.

[b] Good sites: mean windspeed 6–7 m/s at hub height (mid-range in *US Wind Atlas*).

[c] Excellent sites: mean windspeed 7–8 m/s at hub height (top of five classes in *US Wind Atlas*).

[d] Low discount rate is 6%, high rate is 12%.

Sources: The table is derived by the author from data in Cavallo et al., 'Wind energy: technological and economic aspects', in T. B. Johansson et al. (eds), *Renewable Energy* (Washington DC: Island Press, 1993).

1981–5 data from Altamont Pass, California. Recalculated using 1985 exchange rates.

control – could prove important in optimizing wind turbine performance. And materials science, particularly relating to fatigue resistance under high cyclic loads and lightweight materials for rotor and gearbox applications, has obvious applications; corrosion resistance is also important, particularly for coastal and offshore applications.

The impact upon unit costs of the various avenues for technical improvement is hard to assess. Table 4.4, drawing on these developments and the inevitable decline of costs with increased production, shows past and projected costs of wind energy under various conditions. The cost reductions for wind energy since the early 1980s have been spectacular, and further cost reductions may be expected; for average (Class 3) or better sites in Europe (see below), the projected costs post-2000 lie in the range 2.5–5.5 €/kWh, depending on site quality and discount rates. With these reductions, wind energy in many locations appears competitive against new coal, oil or nuclear plant; but even then, at private-sector discount rates, wind energy in most of Europe and elsewhere will still be more costly than new gas turbine plant unless and until gas prices rise, or policies reflect external costs.

4.7 Offshore wind energy developments

Wind turbines can also be sited offshore. Problems of wave loading, salt erosion, access and transmission make it a more difficult and costly environment, but this is offset in part by the much stronger and steadier winds at sea. Some windfarms have already been installed on levées going into the sea, and in shallow protected waters. The first genuinely offshore station, albeit in relatively shallow waters 2.5–5 m deep, was installed in 1991 and is estimated to have generation costs about 40% above those of onshore systems.[21] This has been followed by a 4x500 kW windfarm in the Netherlands, and a 10x500 kW windfarm commissioned in October 1995 at Tuno Knob, off Jutland in Denmark.[22] All are sited in waters of 3–6 m depth a few kilometres offshore. Generally, energy costs are estimated to be 30–60% higher than onshore wind turbines near the same location – though the Tuno Knob plant produced 40% more energy than expected in its first few months of operation. Perhaps the most striking feature of these

---

[21] 'Making history at sea', *Windpower Monthly* (September 1991). The 11x450 kW installation, off a Danish island, is stated to have cost about twice the price of an onshore installation, but after allowing for the stronger offshore winds the energy costs are about 40% higher.
[22] A review of offshore wind activities is given by Gaetano Gaudicioi, 'Offshore wind energy in the world context', *Renewable Enery: Proc. World Renewable Energy Congress*, Denver, Colorado, 1996 (Tarrytown, NY/Kidlington, Oxford, UK: Pergamon/Elsevier Science, 1996).

developments has been their reliability, suggesting that the transition to shallow offshore operation is not as difficult as many feared.

A major feature of the economics of offshore operation is the indication that larger wind turbines would reduce energy costs, since the greater energy output would help offset the relatively fixed costs of foundations, transmission and maintenance visits. In August 1996, the UK utility Power-Gen announced a planning application for the most ambitious project yet, the main proposal being for a 25x1.5 MW installation a few kilometres into the North Sea off the Norfolk coast. Like most major advances in the wind business, this represents a calculated gamble, since the 1.5 MW turbines have not yet been commercially proven for offshore generation. The PowerGen application was just the largest of a flurry of offshore development plans and proposals in the UK and Sweden.[23]

Seabed foundations, however, limit wind turbines to water depths of around 10 m, given present technology. An alternative could be to site wind turbines on or by existing offshore platforms, e.g. retired oil and gas platforms in the North Sea. Could such existing platforms help to provide the foundations and/or mooring that would otherwise be so costly? Bishop, in a commissioned study for this project, found that on-platform wind energy generation was feasible in principle and could be economically attractive for local uses, but that the transmission costs associated with the relatively low power levels would soon become prohibitive.[24] Even if it were possible to install 5x1 MW turbines on a platform, the cost of transmitting the power more than 10–20 km would be prohibitive; at 2–5 km it would roughly double the cost of energy. For abandoned platforms, it would only be attractive for the few cases where loads at such a distance, or grid connections, already exist. Excepting this, Bishop concluded that the most promising applications would be for wind energy to supply power to new or existing platforms, which require up to 10 MW for drilling, production and pumping operations, at relatively high generating costs. This indeed is beginning to emerge; Amoco has installed a wind turbine to provide power on one of its offshore platforms.

---

[23] *Windpower Monthly* (September 1996).

[24] Michael Bishop, 'Economics and policy for the abandonment of oil and gas platforms in the North Sea and an analysis of the potential of re-use for wind energy production', Report to RIIA, 1996.

For deep-water siting without such platforms, at least one design of a semi-submersible structure moored by cables to the seabed in water depths of 30–100 m has been evaluated, but preliminary engineering studies suggest that the uncertainties and costs would be considerably higher than shallow-water seabed siting (even excluding transmission costs), in part because of the high costs of mooring. Such options may possibly be of interest in a broader context of offshore developments sketched in Chapter 7 of this volume.

4.8 European wind energy resources and projections

How much energy could the winds realistically supply in Europe? This is a highly uncertain question for reasons both of economics and the realities of siting. Physically the resources are large. The *European Wind Atlas*, published in 1989, provides estimates of resources at a height of 50 m above the ground throughout the members of the EC (then 12 in number). The maps depict the geographical distribution of five wind energy classes, which can be converted to power densities for different types of terrain (Figure 4.3).

The resources are far from evenly distributed. The UK, Ireland and Greece have exceptional wind resources, which seem likely to offer substantial amounts of economic wind power under many conditions. The southwestern coastal region of France also has some very windy areas. A broad band along the north European coastline, including most of Denmark, and many other coastal sites, and Mediterranean islands, have good wind resources with average windspeed of 7–8 m/s at 50 m height in open farmlands. Much of the rest of northern Europe and the Mediterranean coastal regions have average windspeeds in the range of 6–7 m/s at 50 m. Within each region, of course, local anomalies can give unusually good sites; the *Atlas* is recognized to be of limited accuracy, particularly in mountainous regions, and for example does not capture flow regimes around the Alps that are known to give rise to good local resources in northern Italy.

Converting such data into estimates of potential generation is very problematic. Estimates of gross electrical resources in the first column of Table 4.5 show these to be several times EU electricity consumption. Siting constraints must be imposed on these data. The source sets these as *first-order* constraints based upon specific identifiable exclusions, and *second-*

Figure 4.3 European Union wind energy resources

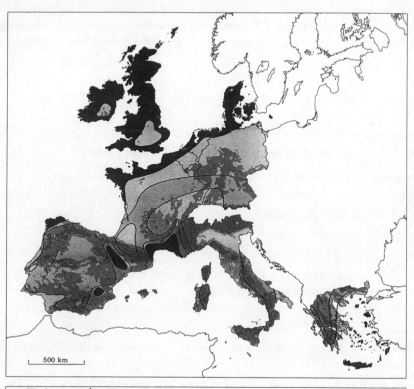

Wind resources[1] at 50 metres above ground level for five different topographic conditions									
Sheltered terrain[2]		Open plain[3]		At a sea coast[4]		Open sea[5]		Hills and ridges[6]	
ms^{-1}	Wm^{-2}	ms^{-1}	Wm^{-2}	ms^{-1}	Wm^{-2}	ms^{-1}	Wm^{-2}	ms^{-1}	Wm^{-2}
> 6.0	> 250	> 7.5	> 500	> 8.5	> 700	> 9.0	> 800	> 11.5	> 1800
5.0-6.0	150-250	6.5-7.5	300-500	7.0-8.5	400-700	8.0-9.0	600-800	10.0-11.5	1200-1800
4.5-5.0	100-150	5.5-6.5	200-300	6.0-7.0	250-400	7.0-8.0	400-600	8.5-10.0	700-1200
3.5-4.5	50-100	4.5-5.5	100-200	5.0-6.0	150-250	5.5-7.0	200-400	7.0- 8.5	400- 700
< 3.5	< 50	< 4.5	< 100	< 5.0	< 150	< 5.5	< 200	< 7.0	< 400

1. Resource densities refer to power in the wind, of which wind turbines can extract 20–30%.
2. Urban districts, forest and farm land with many wind breads (roughness class 3)
3. Open landscapes with few wind breaks (roughness class 1)
4. For uniform wind distribution and land with few wind breaks (class 1); predominance of winds from offshore would increase energy density, winds from land or greater land roughness would decrease it.
5. More than 10km offshore (roughness class 0)
6. Hill siting corresponding to 50% overspeeding, consistent with 400m high hill with 4km diameter base

Source: I.Troen and E.L. Peterson, *European Wind Atlas*, Riso National Laboratory (Roskilde, Denmark: 1989).

Table 4.5 Wind electric potential in selected north European countries

Country or region	Gross electrical resources (TWh/yr)[a]	Population density (/km^2)	First-order potential (TWh/yr)	Second-order potential (TWh/yr)	Electricity consumption, 1990 (TWh/yr)
EC-12[a,b]	8400	140	490 onshore	130 onshore	2000
Denmark	780	120	38 onshore	10 onshore 10 offshore	26
United Kingdom	2600	235	760 onshore	20–150 onshore 200 offshore	285
Netherlands	420	360	16 onshore	2 onshore	67
Norway		13.1	32[c]	12[d]	109
Sweden	540[e]	19		30[f]	140
Finland		14.7	30[g]	10[g]	51

[a] Derived from *European Wind Atlas* (Roskilde: Riso National Laboratory, 1989), including all regions with power density exceeding 25–300 W/m^2 at 50 m height, turbine spacing equivalent to 7 rotor diameters with 26% overall conversion efficiency (see source).
[b] Exclusion factors as for Denmark; see text.
[c] For the Norwegian coast, including small island-cliffs, but none inland.
[d] Using only the best sitings along the coast (ibid. as in note c).
[e] Includes southern Sweden only, and only areas with mean annual wind power densities higher than 450 W/m^2 at 100m height. Offshore sitings at 6–30 m depth and more than 3 km from land are also included.
[f] About 7 TWh/yr at land and 23 TWh/yr offshore.
[g] Including some offshore sitings.
Source: M. Grubb and N. Meyer, 'Wind energy: resources, systems and regional strategies', in T.B. Johansson et al., *Renewable Energy* (Washington DC: Island Press, 1993).

order constraints that reflect an estimate of publicly acceptable levels of installation, based upon experience and very detailed siting and opinion surveys, extrapolated from the few countries for which these are available.

These assumptions lead to a gross electric resource for the EC-12 of about 8400 TWh/yr.[25] The average population density of the EC is close enough to that of Denmark for average exclusions factors similar to those found from detailed Danish evaluations to apply. This would reduce the EC-

[25] See Table 4.5 (note a) for key assumptions. From the *European Wind Atlas*, total land areas with average wind power densities above 250, 400 and 600 W/m^2 are estimated to be 340,000 km^2, 240,000 km^2 and 90,000 km^2 respectively.

12 potential 17-fold to about 490 TWh/yr for first-order land-use exclusions (very close to the technical potential estimated in an assessment for the World Energy Council[26]) and 65-fold to 130 TWh/yr with second-order exclusions[27] – about 25% and 7% respectively of the total EC-12 electricity consumption of 2000 TWh/yr.

More recent estimates of the realistic and publicly acceptable wind energy potential in Denmark suggests that the basis for this figure may be too optimistic.[28] On the other hand, enlargement has added the large, windy and sparsely populated Scandinavian countries which must boost the Union's wind energy resources, and the estimates do not include siting in coastal waters. Overall, a practical onshore and coastal potential in the range of 125–250TWh/yr, or 5–10% of the EU-15's electricity consumption in the late 1990s, appears credible. The latter figure would require land turbines to be sited over an area of perhaps $25,000km^2$, equivalent to under 1% of the Union's land area (and the area between individual machines could still be used for farming and some other uses).

Offshore resources

The constraints on onshore siting of wind turbines naturally lead to greater consideration of the offshore resource. Various national studies highlight various assumptions and results.[29] In 1994 a detailed EC study was completed

---

[26] World Energy Council, *Renewable Energy Resources* (London: WEC, 1994).

[27] An earlier estimate by the European Commission of EC-10 resource *after* quantifiable first-order site exclusions is 4000 TWh/yr. This was based on larger (100 m) machines, included all wind speeds, and used a gridded approach in which sites were excluded only if there were specific identifiable obstacles. The total first-order resource estimated here is also much lower because of the much more severe estimate of constraints when extrapolated from the (much more detailed) Danish studies.

[28] Accumulated experience and renewed debate led in 1996 to a goal of achieving 10% of Danish power supply from wind energy by 2005, along with an indication that this might be the maximum acceptable from onshore wind energy. This is about half the practicable resource previously estimated.

[29] Offshore siting could be important for some European countries if the economics become favourable. A Danish assessment considered only 3 MW wind turbines at sea depths between 6 and 10 m. After accounting for possible conflicts with sailing, fishery, visual impact, wildlife, offshore mining, telecommunications, air traffic and military interests, the assessment

of the offshore potential, taking account of identifiable siting constraints (such as defence areas, sea lanes and marine conservation areas). The results are summarized in Figure 4.4. The estimated technical resource up to depths of 40 m and a distance of 30 km from the land exceeds total EU electricity consumption. The resource is progressively diminished if restricted to shallower sites or closer to shores, to little over one-tenth of this maximum for sites at a depth of less than 10 m within 10 km of the shore. As with earlier onshore studies, however, this reflects only identified constraints and probably over-optimistic technical assumptions.[30] On the other hand, it does not include the considerable offshore resource of Scandinavian countries other than Denmark. Overall, the potential for seabed-based wind turbines may thus be similar to that on land, at about 5–10% of electricity consumption.

These resources are heavily concentrated in the North Sea and off the Atlantic coast, where they are dominated by the UK, Irish, Danish and French potential. In the UK, the potential for machines sited in 5–30 m of water, *after* taking account of shipping lanes, and fishing and military zones, has been estimated to be comparable with total UK electricity demand.[31] It is also clear that Norway has a substantial coastal resource that exceeds any conceivable domestic needs, and recent studies suggest a big potential for siting on Swedish coasts and large inland lakes – though the latter in particular might face strong environmental opposition. For the Scandinavian countries, and perhaps Ireland, extensive exploitation of the offshore resource would be primarily a matter of developing wind power for export to other European countries.

---

indicated sites for about 700 wind turbines with a total yearly production of approximately 5 TWh. It is anticipated that this number may be reduced when actual negotiations with the interested parties, especially the fishery organizations, are undertaken. A number of smaller wind turbines could be accommodated in shallower waters (less than 6 m) where the conflicts with fishery are less serious, resulting in a total offshore resource for Denmark of around 10 TWh/yr. In the UK, the potential for machines sited in 5–30 m of water, *after* taking account of shipping lanes, and fishing and military zones, has been estimated to be about 200 TWh/yr.

[30] Notably, it assumes machines 100 m high with no allowance for array or other losses.

[31] D. Milborrow et al., 'The UK offshore windpower resource', *Proc. 4th International Symposium on Wind Energy Systems, Stockholm* (Cranfield: BHRA, March 1992).

Figure 4.4 Offshore wind electric potential in the EU-12

Annual energy (TWh)

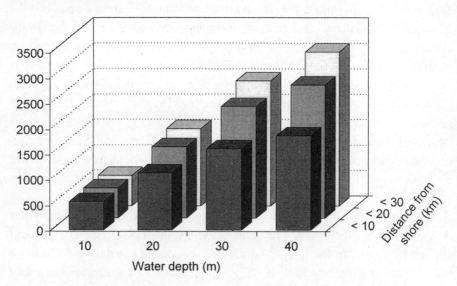

Water depth (m)

Source: H. Matthies et al., 'An assessment of the offshore wind potential in the EC', European Seminar on Offshore Wind in Mediterranean and Other European Seas (Rome: Italian Association of Naval Engineers, 1994).

The theoretical potential for systems moored in deep water is much greater, but, as already noted, this is subject to far greater engineering and cost uncertainties.

Other promising wind resource areas are to be found along the coastal areas of Greenland, Iceland, Finland and the Baltic states, all of which have relatively low population densities.

Thus in Europe, unlike North America, the contribution from onshore wind energy is likely to be limited by the availability of adequate sites; but a long-term contribution of 5–10% of current EU electricity consumption, plus accessible offshore resources of a similar magnitude, nevertheless appears quite feasible.

Timescales of deployment

Over what timescales might these resources be exploited? In 1991, the European Wind Energy Association produced a strategy report, *Wind*

Figure 4.5 Long-term projections for wind energy in Europe

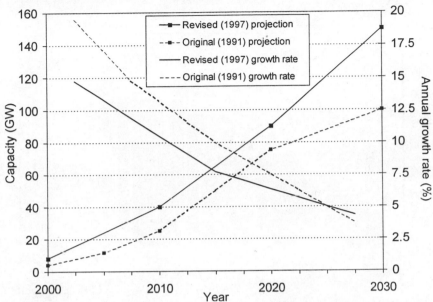

Source: *Wind Energy for Europe: A Strategy Document*, prepared by the European Wind Energy Association for the European Commission Energy Directorate (1991).

Energy for Europe. It stressed the need for 'development of a responsible industry', and 'the need for gentle growth', combined with some sense of where the industry might be heading. To aid the latter, the EWEA proposed goals of 4000 MW in 2000 rising to 100,000 MW by 2030. Six years later – with the target for 2000 already exceeded – the projection for 2000 was doubled to 8000MW and the long-term goal increased to 150,000MW, which would supply more than 10% of EU-15 electricty demand (Figure 4.5).

The percentage growth rate declines steadily and the long-term goal is consistent with estimates of resources given here. Yet even the EWEA's original projections exceed the total projected contribution from all non-biomass sources in the Commission's 'energy vision' scenarios set out in Chapter 1, suggesting that the Commission may be continuing to underestimate substantially the potential contribution of wind energy. The projections suggest that 100,000 MW is treated as the long-term, stable contribution, with new

Figure 4.6 Capacity credit from wind energy as system penetration increases

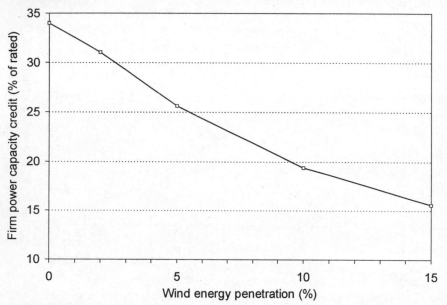

Source: J.S. Holt, D.J. Milborrow, and A. Thorpe, *Assessment of the Impact of Wind Energy on the CEGB System*, Report for the EC DG-XII (1990).

capacity in the period 2020–2030 slowing down considerably, presumably as installations and siting reach maturity. In the context of current negotiations on climate change, it is interesting to note that if the wind energy displaced predominantly existing coal- and oil-fired generation, the projected capacity by 2010 would probably displace over 5% of CO_2 emissions from power generation in Europe, and by the year 2020 potentially two to three times this amount.

4.9 System integration and niche markets in European power systems

The principles of power system operation, and the role of renewable sources in general, have been outlined in Chapter 2. The integration of wind energy in particular has attracted considerable attention because of fears that its apparently random and unpredictable nature would pose great difficulties

Figure 4.7 Probability of extreme output fluctuations for dispersed wind energy in the Netherlands

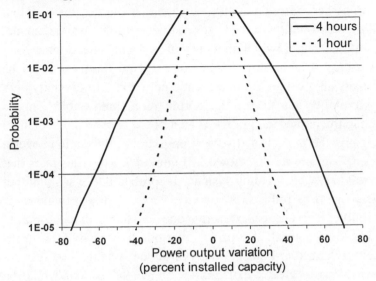

Source: A.J.M. van Wijk, J.P. Coeling and W.C. Turkenburg, 'Modelling wind power production in the Netherlands', *Wind Engineering*, Vol. 14, No. 2 (1990).

for utility operations. In fact this is not the case in most circumstances: electricity demand itself is variable and only partially predictable, and wind energy does not introduce any qualitative difference.[32]

Winds in Europe are generally stronger in the winter, matching well with the seasonal variation in power demand, especially in the north. Short-term variations are more ambiguous. In the UK, within the peak winter periods wind energy and electricity demands appear largely uncorrelated over short intervals. The contribution to system reliability is thus of the statistical nature outlined in Chapter 2. Figure 4.6 shows estimates of the 'capacity credit', as a fraction of the installed wind energy capacity, for the system in

---

[32] See discussion in Chapter 2, and detailed studies reported in F. Walker, 'Integration of new energy in utility networks', *Proc. Euroforum on New Energies* (Paris: UNESCO, 5–6 July 1993).

England and Wales.[33] Given this and the seasonal correlation with demand, the value of electricity from wind turbines probably exceeds that from conventional 'firm power sources' with the same annual average power output at least until the contribution exceeds several per cent of system demand. The value would decline slowly with increasing system penetration.

For wind energy to contribute much power, wind turbines would be spread over many sites connected to the same network. This diversity both reduces its variability and makes the output more predictable. This is apparent even in the limited geographical area of the Netherlands. A Dutch study investigated the impact of diversity on extreme fluctuations over a ten-year period, with the results shown in Figure 4.7. This illustrates that even for the relatively small Dutch system, the probability of wind output increasing from zero to full power, or vice versa, within four hours is entirely negligible. The chances are only one in 10,000 that the output would vary by 30% of installed capacity over one hour or by 60% of installed capacity over a four-hour period. For larger systems – or systems interconnected across a greater expanse – extreme fluctuations would be correspondingly less likely.

Studies indicate in terms of bulk system operation, that wind energy contributions of 20–30% of system demand can be accommodated without having to discard any of the energy and even higher penetrations are possible without technical difficulty.[34] Even on small isolated systems, such high penetration levels appear technically and potentially economically feasible.[35] Fear about the ultimate ability of power systems to accommodate wind energy is one of the greatest myths about wind energy.

---

[33] H.G. Beyer, J. Luther and R. Sterberger-Willums, 'Fluctuations in the combined output from geographically distributed grid-coupled wind energy conversion systems', *Wind Engineering*, Vol. 14, No. 3 (1990).

[34] See in particular the detailed comparison of wind energy modelling studies given in M.J. Grubb, 'The economic value of wind energy at higher power system penetrations: an analysis of models, sensitivities and assumptions', *Wind Engineering*, Vol. 12, No. 1 (1988).

[35] Jens Carsten Hansen and J.O.G. Tande, 'On estimation of the optimal wind energy penetration level', in F.J.L. van Hulle, P.T. Smulders and J.B. Dragt (eds), *Wind Energy: Technology and Implementation*, Proc. EWEA Conference, EWEC 91 (Amsterdam: Elsevier, October 1991).

Table 4.6 Wind energy potential of Greek islands

	Typical windspeeds (m/s)	Energy (TWh/yr)
Cyclades	8.1–10.8	3.15
Crete	8.1	0.74
Euboia	9.2	0.96
Mainland	–	1.61
Total		**6.46**

Source: A.N. Fragoulis, 'Wind energy in Greece — development perspectives and CO_2 reduction potential' (Athens: Centre for Renewable Energy Studies, undated).

The economics of integration at lower levels in power systems – taking account of transmission and distribution constraints and losses – are much more complex. As noted, there are a number of potential niche applications in isolated and small-scale systems in Europe. There are also regions of exceptionally high winds near regions of otherwise quite expensive electricity supply, among which the Greek islands and mountainous regions feature prominently; the islands of the Aegean Sea themselves offer a large high-wind potential that could be connected to the mainland once local demands were satisfied, according to assessments by the Greek Power Corporation (Table 4.6).

As argued in Volume I, it seems reasonable that a greater proportion of the European Union's structural fund expenditures could be devoted to supporting renewable energy in these regions, and wind energy could be a prime beneficiary. If and as special tariff support programmes diminish, these niche markets could help to sustain wind energy in the face of gas generation, but possibly not on a scale much beyond that already established in Europe.

The other potentially wider niche noted in the previous chapter is that arising from capturing the potential benefits of 'embedded generation' in local networks. However, the fluctuating nature of wind energy, the scale of windfarms at several megawatts, and the concentration of wind resources often in regions of low population – and hence low electricity density – limit the scope for this. Indeed, in favoured locations wind energy output is already reaching the capacity of local grids to absorb it, necessitating grid reinforcement to allow export of the power to other parts of the system. Again, therefore, the benefits of distributed generation create a useful niche market but it will hardly provide the foundations to supply 5–10% of European electricity.

For making long-term strategic contributions to European electricity, the predominant technical consideration is that the biggest resources are located around the periphery of Europe. Frequently, by virtue of their low population density, these areas have relatively low electricity demand and most in fact (e.g. Scotland and Norway) already have excess electricity capacity.

As siting constraints in central regions become more significant, attention must turn towards these geographically peripheral areas if wind energy over subsequent decades is to make a large contribution to European electricity supply. Limitations of national transmission capacity, and marginal system losses and costs associated with transmitting power from remote regions through existing or upgraded networks to demand centres, will become increasingly important cost penalties, particularly in view of wind's variability, which reduces the effective utilization of transmission capacity at such high penetrations unless it is combined with hydro-storage. Thus, integration issues may become significant constraints at the local or even national level, even while the European total remains in the range 5–10% of supply.

As part of a long-term strategy for utilizing Europe's impressive wind resources, this issue needs to be given thought. One possibility which deserves analysis would be for DC links through the North Sea to connect Scottish, Norwegian and North Sea wind energy resources to continental Europe, probably landing in Germany. Linking these resources would not only greatly increase the diversity of the wind resources, but also introduce the regulating capacity of Norwegian hydro power. These possibilities are discussed further in Chapter 7.

Between the niche markets of local/embedded generation and long-distance trade from resources on Europe's periphery lies the bulk potential of domestic wind energy production in Europe. This is unlikely either to represent major 'embedded' benefits to the system, or to involve significant operating penalties. It will simply be one of a portfolio of options for domestic power investments. The combination of resources, costs and integration issues discussed above suggests that the likely long-term contribution from onshore and near-shore (shallow water, connected at low voltage to local grids) wind energy will exceed 5% even given somewhat expanded demand, but is unlikely to exceed 10%. Wind energy transmitted long distances from more concentrated remote and offshore resources might ultimately reach similar contributions, but on a longer timescale. The economics and scope for such wind energy in competition for new investment will be determined by regulatory policy towards planning, liberalization, the incorporation of environmental costs and other strategic considerations that will determine the long-term prospects for wind energy as a significant domestic power source for most European countries. To these policy questions we now turn.

4.10 Land-use planning policy

Wind energy development, like any industrial development, is bound to attract some opposition. Whereas many of the environmental objections associated with noise, bird kills etc. appear either greatly exaggerated or resolvable by better design and location, the observation in Volume I that sustainable development is not invisible development applies with most force to wind energy.

Much has been written about the aesthetics of wind energy and public perceptions and attitudes; they are summarized and interpreted extensively in Gipe's book.[36] Gipe argues that some opposition is inevitable but is usually a minority view, and that a great deal can be done to address these concerns and increase the acceptance of wind energy over time.

Specifically, there are two broad challenges. The first concerns that of public perception. If wind energy is widely seen to be a pointless and unpleasant intrusion on the landscape, it will remain a minor power source. If it is widely perceived as an attractive symbol of environmentally sensitive development, the path will be much smoother. The difference depends on public education, siting sensitivities and even cultural dispositions.

The other challenge relates to the design and operation of planning sys-

---

[36] P. Gipe, *Wind Energy Comes of Age*, op. cit. (Note 2). Gipe concludes optimistically that: 'Californians have proved that wind energy can work on a grand scale. Europeans have shown how wind energy can be developed harmoniously with its neighbours and the environment' (p. 482).

tems. The ability to gain planning permission has emerged as a key issue facing wind energy developments in the United States and in the UK, but with very different characteristics. The massive and rapid development of the Californian passes has rooted an image of wind energy as a big, intrusive business in the minds of many Americans, and the planning process has fallen prey to the traditional divisiveness characteristic of much US policy-making. Massive windfarm applications have been opposed by correspondingly massive oppositional campaigns, citing everything from visual rape of the landscape to bird deaths and unreliability of the technology, in a wholesale assault on the proposals and technology itself. Windfarms are proposed and opposed largely on a piecemeal basis, and recovery from the bruised and battered image of wind energy has been hard.

Developments in Europe have tended to be gentler, but planning difficulties and opposition from various quarters have still emerged as an important constraint. In the UK, a national 'Country Guardians' group has led opposition across the country, often with US-style tactics of disinformation campaigns targeted upon national and local media and uncertain local populations. Successes in planning in the second round of the NFFO were down to about 50%, and planning acceptance became the number one issue in the industry. However, a more active response to the attacks, including strong backing for wind energy from the environmental organization Friends of the Earth and from a Parliamentary Welsh Affairs Committee inquiry, together with growing planning experience in local authorities and greater sensitivity of the proposers themselves, seemed to have reversed the trend of rejections by the mid-1990s.[37] In Germany a public backlash, led or supported by the utilities, against the pace of developments (especially in Schleswig-Holstein) has been evident, with the debate in some ways reminiscent of US confrontations. Confrontation with environmental groups, particularly focused on the effect on migratory birds, has also occurred in the rapid Spanish developments.

Nevertheless the dominant pattern of development in Europe has been marked by a wide measure of public support, offset by specific objections

---

[37] House of Commons Welsh Affairs Committee, Second Report, *Wind Energy,* Vol. 1 (London: HMSO, 13 July 1994); Friends of the Earth, *Planning for Wind Power* (London: FOE, March 1995).

(of widely varying vigour and weight) to particular projects. Greater tolerance and support has been widely linked to greater local involvement. In Denmark, community-owned windfarms have consistently found it easier to gain planning permission than those sponsored by the utilities. Elsewhere, the involvement of local communities in the early stages of windfarm planning has proved to be a key factor for success.

Increasingly, therefore, planning approaches in Europe need to orient themselves more towards integrating wind energy as part of local land-use planning, and explicitly as part of local authority development plans, rather than simply waiting to react to proposals from developers. This is already starting to happen. In Denmark, all local authorities are now legally required to develop local plans for the siting of wind turbines in their area. In the UK, since the early 1990s the government's Energy Technology Supply Unit (ETSU) has sponsored regional resource surveys through local utilities and local authorities. In 1993 the UK government issued a Policy Planning Guidance Note that urged local authorities to comment on renewable energy in their wider development plans, stating that:

each authority should consider the contribution their area can make to meeting need on a local, regional and national basis. The contribution should reflect the nature of resources in a particular area and other relevant planning considerations. The planning authority should also bear in mind that investment in renewable energy development can make an important contribution to the national economy, and can help to meet our international commitments on limiting greenhouse gas emissions.[38]

Wind turbines have regained a privileged status in German planning law alongside constructions such as farm buildings, together with measures to return planning powers to local authorities and encouragement for them to develop area-usage plans for wind energy.

Placing the onus more explicitly upon local development plans may impose more obvious constraints on the pace and scale of wind energy development, as compared with letting developers make applications

---

[38] Department of the Environment/Welsh Office, *Renewable Energy*, Planning Policy Guidance Note No. 22 (London: HMSO, 1993).

wherever they wish to pursue them. But the trend towards requiring greater local community and local authority involvement in wind energy developments – combined with acceptance by those authorities of their responsibilities to contribute towards national and international environmental goals, and together with sensitivity exercised by developers – appears to be the key to a more sustainable pattern of wind energy expansion throughout Europe.

4.11 Tariff structures and financing in a liberalized power market

When wind energy is led by private developers, financial viability is, along with planning viability, the other determining criterion. So far, in most of Europe, such developers have been beneficiaries of the curious hybrids noted in Chapter 1: monopoly supply systems with incentives for private generation. Except in the UK and Norway, governments have maintained the basic structure of monopoly utilities, but (for the reasons set out in Volume I, Chapter 2) have regulated them to provide incentives for independent renewable electricity producers – principally by requiring them to pay such producers at some fixed percentage of the sale price of electricity. To put it bluntly, governments have found it not only technically more appropriate but also politically more convenient to support renewable energy by forcing utilities to pay for such support. The justification has been that such payments may reflect costs that the utility avoids elsewhere by accepting such generation; nevertheless, it is essentially forcing utilities to pay for power generation from other operators using sources that they would not themselves have chosen to invest in.

Not surprisingly, several of these utilities have objected vigorously, sometimes with legal challenges. The Danish utility ELSAM, having failed in its domestic political lobbying, appealed to the European Commission in Brussels for a ruling that the obligations contravened European law, but in the summer of 1996 this was again rebuffed. German utilities have been even more vociferous in their efforts, petitioning that the Electricity Feed Law contravenes the German constitution. Again, to date they have been rebuffed, but in some states regulators have resisted enforcing the law and the overall situation remains potentially unstable.

The situation in the UK is different. With the whole system privatized, the major generators have been able to use the NFFO supports to compete in renewable energy on an equal footing, and the system has not been subject to serious challenge by market players. But the NFFO was brought in as a transitional measure and the last round is due in 1998: there is no indication of what – if anything – will replace it and, with the system now appearing on the balance sheets of public expenditure, there is resistance from some Treasury and other government economists to extending the system, which is seen in such quarters as regulatory interference with the idealized free electricity market. Furthermore, the renewable energy industry is itself concerned at the bureaucratic and costly nature of the selection process. Both difficulties may grow as the scale of activity increases.

Against this backdrop the juggernaut of European electricity liberalization is moving. As sketched in Chapter 1, this will evolve in different forms in different countries, but increasing competition in power generation and consumer choice is clearly on the way. This, combined with the expanding scale of wind energy activities, will have profound implications for the future of wind energy.

As explained in Chapter 1, liberalization represents both a huge opportunity for renewable sources and a huge challenge. Both may be focused most of all upon wind energy. At the same time as creating a market structure in which independent power generation is a formalized and integral part of the electricity business, liberalization will vastly improve the scope for gas-fired power generation, and increase pressure on the current support regimes. Utilities in continental Europe may complain about the current situation, but at present they can at least pass the costs of the incentive schemes through to their captive customer base. If they start losing industrial consumers to private, gas-based generation and companies that do not bear the costs of the renewable energy tariffs, their complaints will carry far more force, and the fiscal base of supports will be steadily eroded.

Some elements of current support regimes do not hinge upon electricity industry structure and tariffs. Given the relatively capital-intensive nature of wind energy compared with gas-fired generation, the cost of finance will be of enduring importance. An additional part of the system in Germany is the availability of relatively low-cost finance through the Deutsche

Ausgleichbank (DAB), which is part of the Ministry of Economics. It provides low-interest loans that cascade through the banking system to cover up to 75% of the capital cost of wind energy developments at interest rates lower than would otherwise be available for such investments (7.5% in early 1993). More generally, the differing financial communities across Europe result in different kinds of finance, with the traditional short-term nature of finance in Britain increasing the cost of capital there.

Other issues relate more explicitly to the incorporation of 'external costs'. Whereas this is a primary justification of other support mechanisms, in some countries there is a more explicit link; Denmark is one of the few countries to have carbon, nitrogen and sulphur emission taxes, from which wind energy is exempt. The ability of governments to introduce such mechanisms will be of particular importance if other supports are eroded through liberalization.

The options for the future involve issues generic to the support of renewable energy, and are accordingly considered elsewhere. The point about wind energy is that, as the most successful, most developed and most visible of the new renewable electricity sources, it will be the first in the firing line of these changes, and potentially best-placed to influence them. And that in itself will be a real test of the ability, strength and sophistication of the industry.

4.12 Prospects and policies for the wind energy industry in Europe

The development of wind industry in Europe forms a remarkable story. In the space of little over ten years from 1975 it rose from a small, largely craft- and workshop-based effort of enthusiasts into an international industry, an industry that has attracted more than 4 billion ecus of investment and is now one of Denmark's largest foreign currency earners. Danish manufacturers play the major role in Europe and the international market. German, Dutch and British companies have arisen to supply domestic markets and compete internationally, and several other European countries have companies that have arisen particularly for the domestic market.

Wind energy is one of the few rapidly growing global engineering industries in which Europe has an unquestionable lead. The much-hyped initiatives by US manufacturers in the early 1990s have come to little, given the wide-ranging difficulties of the US market: Kenetech, which had sought to lead the world with its variable-speed machine mass-produced for big windfarms, struggled with technical, financial and political problems in the difficult US conditions and finally sank into bankruptcy during 1996. Mitsubishi, too, which focused primarily on the US market, has been caught in the morass, though it is turning towards Asian and European markets. The tragedy of wind energy in the United States is Europe's opportunity.

Nevertheless, all is not well. Most countries except the UK have sought to foster a domestic industry by preferential treatment in the home market. As a result of this and the rapid expansion, the European industry is fragmented, with a proliferation of small companies alongside the half-dozen or so 'majors' – companies that are themselves still tiny in comparison with the national engineering and energy companies. A curious feature is the absence of any of the big engineering companies, apart from Taylor Woodrow's backing for the Wind Energy Group plc. This seems particularly curious in Germany, where the major engineering companies either appear to express open scorn, or seem happy to remain as component suppliers – though Siemens has sought to capitalize on the latter role with an announced aim of setting an industry standard with a DC 750 kW generator.[39] May 1996 finally saw the first major change in this situation, when the Deutsche Babcock subsidiary Balcke-Durr bought a majority holding in Nordex wind turbines, to focus on production of a 1 MW machine.[40]

To maintain a global lead against competition arising in Asia, the European industry may need companies of greater financial strength. This needs to be based on some rationalization of the industry, financial flows derived from a steady market expansion and the backing of bigger players, and continuing growth of technical sophistication and market experience. The consequent issues for public policy can be cast in terms of the five policy challenges set out in Volume I (Summary and Conclusions).

First, there is clearly a continuing role for RTD and associated education and dissemination programmes. European programmes will continue to

---

[39] 'Giants wary of involvement', *Windpower Monthly*, Vol. 11, No. 9 (September 1995).
[40] 'Industrial giant buys into wind production', *Windpower Monthly*, Vol. 12, No. 5 (May 1996).

form the backbone of this; most of the major advances in machines and new applications have involved support from the EC's JOULE-THERMIE programme, and there is still much scope for advances in the directions outlined earlier in this chapter. Wind energy RTD appears to have successfully avoided many of the pitfalls discussed in Volume I, Chapter 6, and a major public scandal over the diversion of funds away from renewable (particularly wind) energy projects in 1995 appears to have been resolved with greater transparency of procedures and an additional round of tenders,[41] suggesting that the political base of support for these programmes will not easily be weakened.

Second, although the protection of national industries in many European countries has had a useful role in helping to foster a domestic constituency and a variety of technical paths, it has now outlived its usefulness. The normal rules of the Public Procurement Directive should be applied to purchases by public utility companies, and – as in the UK – supports for independent generators should be open to all manufacturers. This should include foreign manufacturers, not only because of basic trade principles, but more particularly because the European companies will be able to compete in the huge emerging Asian market only if they can first compete successfully against Asian manufacturers in Europe. Supports for wind energy are not themselves under any near-term threat from international trade regimes, since wind energy is clearly not a 'like product' to other energy sources (though this question could start to arise in the case of major electricity trade). But the provisions of the WTO and the Energy Charter Treaty should apply to competition between different manufacturers, and the wind industry should at least keep an eye on the daughter Charter Treaty negotiations intended to develop a regime for new energy investments (for a brief outline see Volume I, Chapter 5, and references therein).

The third requirement is to ensure steady expansion of the European market. Some of the general experience with market-building has been described in Volume I, Chapter 7, and specific aspects of the supports for wind energy to date have been set out above. Providing for market confidence will require European governments to set credible national goals

---

[41] 'Another bite at the budget', *Windpower Monthly*, Vol. 12, No. 1 (January 1996).

well beyond the year 2000, building to a European goal. Such targets could usefully be developed in the context of negotiations on the new round of climate change commitments. They would serve as a guide for the future development of tariff and financial regimes and their transformation in the course of electricity privatization.

That transformation provides the fourth major challenge. If European electricity systems proceed towards fuller competition, it is unlikely that the general diverse investment projections summarized in Chapter 1 will be realized. Wind energy can survive a 'dash-for-gas' in the context of a continuing diversity of investments, but not a *Blitzkreig* in which wind energy could be as much a victim as conventional major thermal plant. Should this situation emerge, measures would be required to ensure that investors could capture some of the value of diversity as well as environmental protection. More generally this reflects the importance of implementing systems to reflect external costs in the prices of the various energy sources, discussed in Chapter 8.

The final challenge is that of integrating wind energy with other policy agendas. As noted, wind energy resources are good in all the less-developed (cohesion) countries, and the wind energy industry is at the stage where it could make an important contribution to development of the electricity sector in all these countries; the use of structural funds should reflect this in the context of Europe-wide commitments on climate change. Some consideration could also be given to wind energy in the context of offshore oil and gas platforms (Chapter 7). Wind energy needs to be integrated into local land-use planning across Europe, harnessing also the goodwill derived from local Agenda 21 efforts. And, perhaps above all, given the possibly explosive development of Asian markets, wind energy needs to be integrated in the drive to maintain international competitiveness and to build foreign markets in the mechanical and electrical engineering industries, through trade missions and the increasing scope for promoting environmentally clean technologies by means of bilateral and multilateral aid and mechanisms such as Joint Implementation under the climate negotiations.

To conclude, all the evidence set out in this chapter suggests that wind energy has the potential considerably to exceed the projections set out in the Commission's most recent Energy Vision projections, and that it could become a major global source of clean energy, in which European companies maintain a global lead. The record of European policy to date, despite some hiccups, has been good. The determining factor now will be how well the industry and policy-makers together manage the transition to a global industry in increasingly liberalized electricity markets.

Chapter 5

Solar Electricity

The solar energy falling on Europe amounts to several hundred times European energy consumption. Electricity from solar energy, particularly photovoltaic cells (PV), has long attracted particular attention among the renewables and a large share of both public and private sector RTD. However, PV use accounts for under 0.005% of European electricity supply, because it is 5–10 times as expensive as conventional power generation. A market exists because of PV's application to many special, small applications, and because of installation programmes funded directly by government.

However, the small size of the current market implies great scope for reducing costs with the scaling up of production, and many technical options have been identified for additional cost reductions. Nevertheless an economically sustainable market must also fully exploit the potential value of PV's special characteristics, notably the very small unit size and lack of almost any local impact during operation. Crystalline cell technology dominates European manufacturing and RTD policies need to explore new applications, but costs are unlikely to fall below 25 ∈/kWh by 2005, which is the European Commission's objective, and silicon supplies may limit ultimate applications. Industrial pressures for supporting PV applications based on existing technologies should thus not be allowed to eclipse RTD into the thin-film technologies that offer the greatest promise for major cell cost and material reductions.

PV is ideal as a component in 'distributed resource' systems, for grid support and generation at the point of end-use. In general the economics will improve the lower the level in the system, and the greater the integration of PV with other structures that offset balance-of-system costs. This points to siting on building surfaces – rooftops, and integrated as cladding – as being a primary application in Europe, perhaps with limited on-site storage. Such applications will be enhanced if local DC systems evolve, reducing costs and conversion losses in both generation and end-use equipment. The technical resource for rooftop applications is 25–40% of total electricity demand in Mediterranean countries and 12–18% in central and northern Europe, declining below 10% in Scandinavia. Applications to service-sector buildings, perhaps in combination with co-generation in northern Europe, may be particularly promising. Electricity liberalization reaching the level of individual end-users would help such markets to emerge.

In principle, some applications to transport could attract similar benefits of integration, for marine uses and with siting alongside railways and, for charging electric vehicles, motorway verges and car parks. The synergies at present are speculative and markets would be inextricably tied to the use of non-petroleum vehicles, which for roads is negligible at present.

Industry targets call for PV to supply more than 0.1% of European electricity demand by 2010. Foreseeable premium applications for PV in Europe (principally buildings) may amount to a few per cent of electricity supply. In total, however, dispersed energy uses – for which PV could in principle have some comparative advantage over centralized supplies – amount to perhaps half of European energy demand. The huge gap between this and specific identified opportunities can only be narrowed by as yet unforeseen innovations in solar technologies and European energy systems. The extent of innovations required means that expansion of PV, despite its greater ultimate potential, is likely to be both later and slower than wind energy; contributions by 2030 are likely to be in the range 2–6% of European electricity supply. However, because its applications would be associated with higher cost and more efficient electricity uses, its overall economic and environmental value would be disproportionately greater.

By far the greatest potential for PV lies in the developing world, where more than a billion people remain without power. Building on relative success in a few countries, PV is beginning to take root as a reliable and credible option for supplying power to remote homes and villages. However, funding remains a major constraint and aid organizations have not been convinced that PV should be backed as an efficient route to rural development, poverty alleviation or sustainability. Accelerating such applications may, however, also be justified in terms of the cost reductions bound to follow from expanding PV production, and recent initiatives led by the World Bank may succeed in transforming the market.

European companies have some advanced technologies but not an all-round competitive advantage. Fostering innovation – in technologies, applications, systems and institutions – needs to be the guiding principle of policies. The need to develop better all-round techniques for European applications could be a driving force for global competitiveness, and policies should not undermine that pressure simply by subsidizing easier and more conventional applications. Components of policy could comprise:

● continuing RTD, focusing particularly upon innovative applications (especially for buildings and transport) and thin-film technologies;
● pursuit of electricity liberalization combined with support for PV-related end-use equipment (e.g. DC networks and appliances and two-way metering) and educational programmes;
● pursuit of greater cost transparency in the electricity industries, including internalization of local environmental costs particularly in electricity distribution, and in some transport applications;
● development of cost-reflective electricity pricing and buy-back rates, or equivalent premium payments for self-generation;
● integration with other Union policies, particularly relating to Mediterranean development, international aid, and export strategies.

5.1 Introduction

Solar energy is the ultimate source of most renewable energy, and solar radiation at the Earth's surface represents by far the largest renewable energy resource. In Europe annual insolation declines from a peak of almost 1,800 kWh/m^2 in the extreme south to about half of that level in the north of Britain and Scandinavia (Figure 5.1). This compares with desert insolation that can reach almost 2,500 kWh/m^2. Nevertheless the annual solar energy falling on Europe is several hundred times Europe's annual commercial energy consumption. Such a huge resource inevitably attracts attention and policy effort, particularly since solar electricity production has less visual or other local environmental impact than most other renewables.

One approach to generating solar electricity is with thermal solar power

Figure 5.1 Solar radiation in Europe: annual insolation (kWh/m² per year)

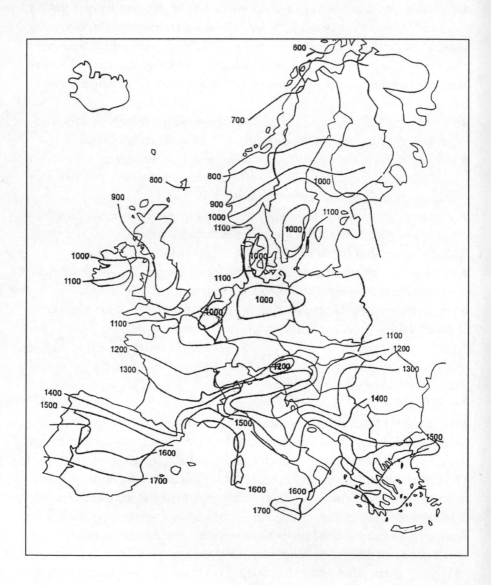

Source: European Commission, *Photovoltaics in 2010*, EC DG-XVII (1996).

technologies, in which solar radiation is used to raise steam for power generation. As outlined in Box 5.1, these technologies have developed rapidly. Nevertheless, their dependence upon sustained, direct, clear-sky, solar radiation, combined with the fact that they are centralized power sources that cannot take advantage of the potential system benefits discussed below for photovoltaics, makes it unlikely that they can compete against conventional power sources or even other renewables in most European conditions. Solar thermal power technologies could prove more attractive on the other side of the Mediterranean in North Africa, and as outlined in Chapter 7 this could have an important European dimension. Therefore, this chapter focuses upon the 'primary' solar electric technology: the generation of electricity using solar photovoltaics (PV).

Solar photovoltaic devices convert light directly into electricity using a method that differs fundamentally from all other modes of electricity generation. A PV generator is made of small individual cells incorporated in a module encapsulated in glass or plastic. Such modules are typically designed to give around 50–100 Wp[1] direct-current output at around 12 V. To obtain higher power, the modules are connected together in an array of the desired size, oriented to collect maximum sunlight.

At present, PV electricity is more expensive than that from conventional bulk electricity sources, or even from wind energy and several other renewable electricity sources. Nevertheless there is great interest in photovoltaics. There are widespread applications (such as for telecommunications and consumer products) arising from the convenient, simple and modular characteristics of photovoltaics. A commercial market already exists, based on such applications and increasingly on various grid-connected applications; after a brief hiatus in the early 1990s, global PV sales resumed expansion and in 1995 grew by 14% to exceed 72 MWp/yr (72 MW at peak output).

Costs have been falling steadily since the beginning of terrestrial applications and meanwhile other markets are expanding; the fall is expected to continue because of the many options available for further development, as will be described below.

---

[1] Wp is peak Watts, the maximum output of the PV system.

Box 5.1 Solar thermal power technologies

Four main ways have been explored for thermal conversion of solar energy to electrical power.

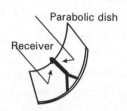

- **Parabolic trough plants** have a linear parabolic reflective surface which concentrates direct sunlight onto a receiver tube which runs along the focal line. Fluid in the tube is heated and transported to a central generating boiler. Auxiliary gas heating can be used for the same boiler to ensure reliable output during cloudy weather and (if appropriate) at night.

- **Parabolic dish plants** concentrate direct sunlight onto a receiver at the focal point of a parabolic dish which is oriented by two-axis tracking to point directly towards the sun. Some receivers generate electricity directly; others collect the heat in a working fluid which is distributed to a central boiler and generator.

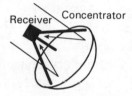

- **Central receiver systems** consist of a field of heliostats, or sun-tracking mirrors, which reflect direct sunlight to a tower-mounted receiver. Fluid in the receiver absorbs the concentrated energy and transfers it to a boiler for power generation.

- **Solar ponds** are ponds with a high gradient of increasing salt concentration towards the bottom which prevents hot water rising. Consequently the pond traps solar radiation, building up temperature near the bottom which can then be transferred for power generation.

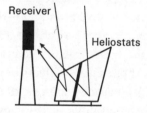

The first three systems rely upon collecting direct sunlight; the last also uses indirect sunlight, but its application is limited by relative low efficiency and very high water requirement (it uses about 35 times as much water as a conventional power plant). Experience with central receivers in Europe has been disappointing, and parabolic dish systems have proved expensive because of the need for good optics, two-axis tracking and the limited scope for economies of scale.

Only parabolic trough plants have seen significant commercial development. Between 1984 and 1990, Luz International built nine with a total of 355 MWe in the Los Angeles region. As with wind energy developments, the first installations were relatively high-cost and low-efficiency, but the systems were scaled up and improved radically. The final two plants, SEGS–VIII and SEGS–IX, had rated capacity of 80 MW each with field areas of around 464,000 m^2 and 484,000 m^2, and generate around 255 GWh/yr at an estimated cost of 8.9 ¢/kWh.[a] In these conditions, about 2 ha of land are required per megawatt of installed capacity. Projections by Luz suggested a cost of 8 ¢/kWh for a 200 MWe plant in the Mojave Desert, and independent assessments suggest 8–12 ¢/kWh.[b] But before such plans could be realized, Luz went bankrupt, a victim of the same forces which halted the Californian wind industry.[c]

Box 5.1 *(cont.)*

Despite the impressive development of parabolic troughs particularly, solar thermal power seems unlikely to be a technology of major interest in Europe. Given the lesser solar resources, a duplicate of the advanced 200 MW proposed plant, assessed at a 6% real discount rate and placed in the deserts of southern Spain or Italy, seems unlikely to generate power at less than 0.1 ecus/kWh. Central receiver stations are calculated to be of similar or lower cost than parabolic troughs, but seem even less appropriate to Europe because of their high sensitivity to interruption of sunlight.

Although these costs are much cheaper than PV power at present, the scope for cost reductions seems more limited, and they do not have the unique small-scale and modular characteristics of PV. The efforts to capture solar thermal power have led inexorably to large-scale centralized technologies which must compete directly against conventional power generation. This rules out the kind of niche applications through which PV may develop, and also raises concerns about finding acceptable sites, again a particular drawback in Europe. Consequently, although solar thermal power is of potentially great interest in the southern United States and many developing countries, it is not discussed further here, but is considered in the context of Mediterranean developments in Chapter 7.

[a] 6% real discount rate used throughout.
[b] Pascal de Laquil III et al., 'Solar thermal electric technology', in T.B. Johansson et al. (eds), *Renewable Energy: Sources for Fuels and Electricity* (Washington DC: Island Press, 1993).
[c] A concise account of the rise and fall of Luz International, and its technology, is given in 'Luz: Final Point or Eclipse?', *Best of Systems Solaire No.1*, Société d'Information sur les Energies Renouvelables (Paris: SARL, 1993).

5.2 The characteristics of photovoltaic energy

Photovoltaic generation differs radically in its nature from almost any other source in three ways. First, the primary solar resource has a high overall conversion efficiency – at present around 10% for the overall system and likely to improve to 15–20%. Thus, relatively small areas, such as building surfaces, can generate useful amounts of power, and the theoretical resource available from photovoltaics is much larger than from any other renewable technology. Secondly, as noted above, the intrinsic unit size is very small, thus allowing photovoltaics to be used in many small and varied applications. Because of this they can have an advantage over many other sources in non-grid markets and a variety of other applications. Thirdly, a plant has little or no visual or environmental impact during its operation.

Being so fundamentally different from fossil-fuel-based and even other renewable energy sources, photovoltaics offer peculiar benefits and problems.

Basic intrinsic favourable characteristics are modularity, simplicity, no fuel requirement and an absence of noise, wastes and emissions. Modularity means that plants can be located close to the users, thus reducing transmission and distribution costs, and when required can be installed at short notice. It also means relative ease in finding sites for photovoltaic arrays, even in urban areas (rooftops, façades, car parks etc.), the only requirement being access to sunlight. Simplicity, due to the absence of moving parts, implies high reliability and low maintenance requirements.

The main difficulties of photovoltaics concern (a) costs and (b) the fact that the electricity is generated only in daytime. For remote or isolated applications this usually necessitates adding storage devices to the system. For grid-connected applications the value of the electricity depends on how sunshine correlates with electric system demand (e.g. photovoltaic power is especially valuable in southern regions where air-conditioning systems create peak demands during hot summer days, and in office buildings with peak midday consumption).

No environmental pollution is created when plants operate; however, environmental and safety hazards can occur during manufacturing. The risk, common to other solid-state electronic production, comes from the use of hazardous materials (solvents, hydrogen selenite, cadmium etc.). As the industry expands, such impacts must be reduced to low levels by adopting modern waste-minimization and recycling techniques.

Other components are needed along with the module to make a complete plant. These 'balance-of-system' (BOS) components include support structures, electrical system components (cables, breakers etc.), inverters and batteries. Therefore to achieve overall photovoltaic system performance improvement and cost reduction requires coordinated effort. Even if some components (e.g. support structures, batteries) seem to have already reached their technological maturity with a low margin for further improvements, there are others (e.g. inverters) where there are good prospects for obtaining higher quality, improved reliability and cheaper components.

5.3 Photovoltaic energy developments

Although discovery of the photovoltaic effect dates back to 1839, it was only in the second half of the twentieth century that scientific and technological development produced photovoltaic cells for practical applications.

The first application was in space satellites: from the Vanguard (1958) to today's Skylab, photovoltaics became the conventional power source in space and are likely to remain so for the foreseeable future. Photovoltaic devices for such applications were very dependable, even in the very difficult environmental conditions of space. In addition, their high cost was acceptable, being only a small fraction of the whole cost of space projects. This development gave rise to a small RTD-based industry which supported the growth of a new market segment, namely the 'service' consumer applications: calculators, watches, toys and other gadgets.

The next milestone was the energy crisis in the early 1970s that made it possible for the first time to consider photovoltaics as an energy supply option despite the huge costs. Governments embarked on major RTD efforts to exploit the possibility of developing less expensive photovoltaic devices suitable for terrestrial applications. From the 1970s, PV enjoyed a favoured position in many government RTD budgets, and also attracted the backing of some major energy companies. As a result, new industries and new growing markets arose. From then on, even with alternating hopes and disappointments, the photovoltaic systems market grew together with a continuous increase in conversion efficiencies (see Figure 5.2). Almost constant nominal prices since the late 1980s reflect slowly declining real (inflation-corrected) prices in the relatively stagnant market; module prices in the mid-1990s appear to have started declining more rapidly again (Figure 5.2).

Today photovoltaic power systems are already cost-effective and commonly employed in a wide range of remote applications – for example, electricity supply to isolated users and small communities; water pumping and desalination; powering of service equipment such as radio repeaters, pipelines and well-heads; and cathodic protection. Such remote applications accounted for 83% of total PV sales from 1990 to 1994 (Figure 5.3). A promising new market is emerging for grid-connected applications, which are projected to grow from 11% in 1990–94 to 29% in 2010 of the total market according to these estimates.

Figure 5.2 Photovoltaic power modules: shipments, average selling prices and module efficiencies for 1980–95

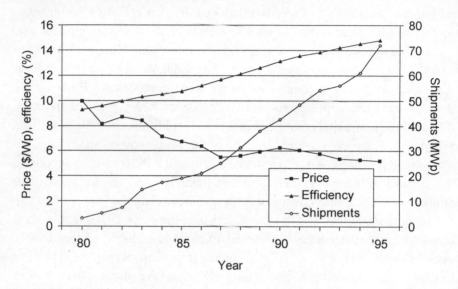

Source: ENEL, RTD department.

Governments and institutions in the main industrialized countries have launched very important demonstration programmes. In addition, especially in developing countries, photovoltaic decentralized energy production and supply offer pre-electrification or even basic electrification (village power) and may reduce urbanization through enhancing the quality of life of rural populations.

However, in the early 1990s, the PV industry was faced with low oil prices, a world-wide economic recession and costs which had not declined as fast as was hoped. This resulted, at the very least, in delays in many of the larger photovoltaic programmes of many of the companies. Production capacity has exceeded market demand, and revenues are insufficient to self-finance the RTD efforts still needed in order to achieve the breakthrough necessary for photovoltaics to become a significant energy source. Faltering industrial confidence and investment increased the need for development efforts that pool resources and use them in the most efficient way.

Figure 5.3 Market share by major applications

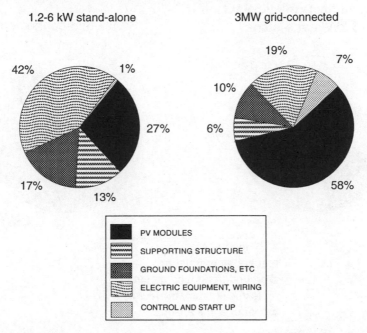

1.2-6 kW stand-alone

3MW grid-connected

PV MODULES

SUPPORTING STRUCTURE

GROUND FOUNDATIONS, ETC

ELECTRIC EQUIPMENT, WIRING

CONTROL AND START UP

5.4 Cost of photovoltaic electricity

Despite a decrease in the price of modules (see Figure 5.2), the cost of electricity from PV still remains much greater than that of power from traditional power plants. The cost of electricity from photovoltaics basically depends on the amount of sunlight available at the site; the cost of modules and BOS components; the cost of building, operating and maintaining the plant; and (last but not least in such a heavily capital-intensive source) the cost of money and the plant lifetime.

The way each component contributes to the final energy cost also differs for the various applications. For stand-alone systems the storage, usually lead-acid stationary batteries, contributes some 12–16% to the overall cost. In remote areas the transport and mounting expenses can become relevant. Installation costs vary according to whether a photovoltaic array is sited on the ground or whether existing structures, such as rooftops, are used and, mainly in large plants, whether some pre-mounting operations are performed at the factory or whether all operations are carried out in the field.

Figure 5.4 Cost breakdown for small stand-alone and large grid-connected photovoltaic plants in Italy

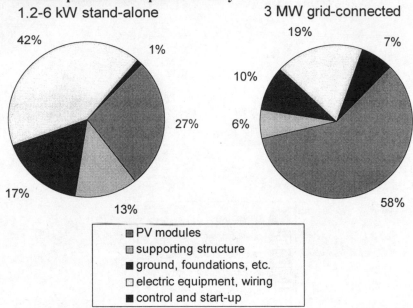

Source: ENEL, RTD Department.

As an indication of how the cost breakdown can vary, Figure 5.4 shows the data from two different operations carried out by the Italian utility ENEL: the construction of stand-alone systems for 'wireless electric service' to remote users and the large (3 MWp) grid-connected plant (completed in 1994) in Serre, south of Naples.

On the basis of the experience acquired, some cost ranges can be given (see Figure 5.5) for the various applications. Whilst there are economies of scale for larger applications, these are not as great as the corresponding decline in costs of alternatives against which PV would be competing at the different scales. Whereas for remote applications photovoltaics can already be a cost-effective option, it is harder to see how they can compete with energy from the grid. In fact, in remote areas PV may be competitive with traditional small diesel generators, or with grid extension just a few kilometres away. The main obstacle to a larger use of photovoltaics, especially

Figure 5.5 Cost of power produced by different PV systems

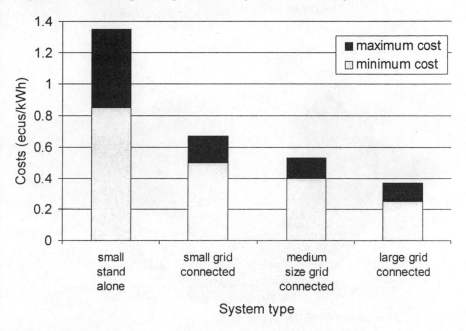

Note: Lifetime 25 years, discount rate 8%, 1700 equivalent hours insolation.
Source: ENEL, RTD Department.

in developing countries, derives from the need for an up-front high capital investment. When compared with the cost of electricity delivered from the best modern conventional power stations, photovoltaics at present appear five to ten times more expensive. This is too large a gap to be closed solely by incorporating environmental costs and distributed benefits (see Chapters 1 and 2). Large-scale application of PV thus hinges, first and foremost, upon substantial cost reductions in both cell technologies and complete systems. However, one important feature of PV is the fact that, owing to the small current market and various economies of scale in production, large cost reductions can be expected simply from scaling up operations.

So throughout the early 1990s, the industry and observers eagerly awaited the first big, bold step. In 1995, the international gas company Enron announced a joint venture with Solarex to build a new facility to produce 10 MWp/yr of amorphous silicon cells, an order-of-magnitude increase

Figure 5.6 Share of different technologies in the European industry

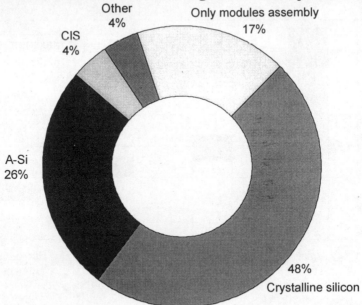

Source: Photovoltaics in 2010, EC DG-XVII, 1996.

compared with previous scales. Much of the output was intended to be used to construct a solar power plant in India. It was claimed that, aided by some relatively low-interest loans and a 0.75 ¢/Wp support from the State of Virginia, the fully installed cost would be about $2500/Wp, and that this would produce electricity at an estimated cost of only 7 ¢/kWh. At the time of writing, the production plant is operating but at a much lower output of around 3–4 MWp/yr, and the Indian contract has yet to be signed; it is still expected to proceed, but with a mix including more conventional crystalline cells.

5.5 Potential technology developments and RTD

Currently wafer-based crystalline silicon technologies account for about half of European PV industry production and amorphous silicon for another quarter (Figure 5.6). To achieve long-term major cost reductions, strong RTD efforts are required on a 'twin track':

- to continue to enhance performances of crystalline silicon technology to its physical limits, including the balance-of-system (BOS) components and the whole-system design. In this manner larger market segments will be opened up, experience with siting finance and markets will be developed, and the social acceptance of photovoltaic technology will be increased.
- to foster new and ultimately more promising technologies for the future, taking into account the time needed to develop it at the appropriate industrial level, transferring present laboratory findings into a steady commercial production.

Crystalline silicon technologies

Crystalline silicon technology has the advantage of using a largely available, non-toxic, well-understood base material. In addition it has a well-proven reliability and a high energy-conversion efficiency (at laboratory level almost the theoretical maximum value has been reached). Transferring such laboratory results into cost-effective industrial processes will lead to further increases in the efficiency of commercial products.

However, some factors still hamper such technology. The relatively high amount of crystalline silicon required contributes substantially to final cost. At present the technology requires about 20 tonnes of silicon per megawatt peak output, and derives its feedstock from the off-grade product of the solid-state electronic industry (at present about 700 tonnes/yr of silicon scraps are usable by the photovoltaic industry).

Although the electronic industry capacity can satisfy near-term supplies to the PV industry, in the longer term its growth will typically be below that of PV; in view of this, many efforts have been devoted to developing a chemical industrial process to produce 'solar grade' silicon, and European companies are leaders in this field. Also, the demand for both material and cost could be reduced by techniques that allow material to be used more efficiently, principally by improving the ingot-cutting techniques to obtain ever thinner wafers. However, as the thickness of the wafers decreases, their fragility increases, so particular care in handling them is needed during all the production phases. In addition, reducing thickness beyond certain limits reduces efficiency and may require expensive additional steps to compensate for this.

Figure 5.7 Thin-film module efficiencies

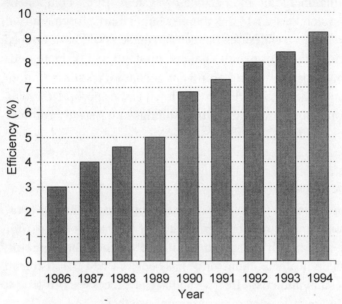

Source: R.C. Kelly, Amoco/Enron Solar, 'The emerging solar power market', presentation to Montreux Conference (16 June 1995).

A drastic solution would be to avoid sawing by producing thin silicon ribbons, suitable for immediate processing to make cells. Several RTD efforts in 'ribbon growth' technologies have had some good technical results but no immediate prospects for commercial applications, because of their low productivity.[2] A similar situation is being found in other approaches, as for example the Texas Instruments 'spherical' cells matrix.

Thin-film technologies

In principle, the technology that could produce low-cost, highly efficient and highly reliable photovoltaic devices is based on thin films. These are made of exceedingly thin layers of semiconductor material that require very

---

[2] K. Zweibel, *Harnessing Solar Power – The Photo-voltaic Challenge* (New York: Plenum Press, 1990).

little material and do not need to be grown or sliced. Entire modules may be built as a single unit; layer upon layer is deposited on a substrate. For a long time amorphous silicon thin film has been regarded as the most promising technology owing to its low use of silicon and the possibility of a low-cost process and integrated production (i.e. large devices are directly produced with no need for multi-step processes). Thin films also benefit from RTD efforts in the electronics industry. Unfortunately, the results achieved more than 20 years after the early announcements do not fully confirm the expectations. Production lines exist but the product is still more expensive than crystalline cells, and has a poor efficiency which, moreover, declines with exposure to sunlight, though there have been big improvements (Figure 5.7).

Other materials are being investigated to produce thin-film photovoltaic devices. The most promising are copper-indium diselenide (CIS) and cadmium telluride. Such materials are more stable than amorphous silicon but they have exhibited some problems when laboratory results are transferred into actual products owing to the complexity of the devices and related manufacturing processes. In addition, toxic materials are used during their production.[3]

Different cell structures are also studied in order to maximize efficiency. In this field multifunction (tandem) cells seem to be very promising. Such devices allow for high conversion efficiency using the whole spectrum of radiation energy by stacking elementary cells made from materials that absorb different radiation bands.

Manufacturing and balance-of-system technologies

Factory automatization will improve the manufacturing yield by reducing the number of cells lost, notably towards the end of the process where the cost accumulated in the product reaches its highest volume. That means that the photovoltaic industry should aim at becoming a 'zero-defect' industry, requiring – from an early stage – appropriate actions for quality control.

---

[3] P.D. Moskowitz et al., 'Toxic materials released from PV modules during fires: health risks', *Solar Cells* (1990); L.D. Hamilton et al., 'Public health issues in PV energy systems: an overview of concerns', *Solar Cells* (1987).

Such actions will be especially important if, owing to growing production rates, specialized manufacturing steps shared among different module producers become the rule.

The need for specific RTD actions concerning balance-of-system components will acquire an ever greater importance as the cost of photovoltaic modules decreases, so as to limit the BOS contribution to overall cost (already around 50%). Since such components use better-known and more widely diffused technologies, the margins for performance improvements and cost reduction seem lower than those for modules. However, possibilities do exist with regard to the appropriate integration of PV elements into building structure; the design of highly efficient, reliable, safe inverters; the design of light, low-cost, support structures; and the definition of simplified cabling and installation methods. As photovoltaic grid-connected applications develop, it will be increasingly important to understand and to solve the problems related to grid interfaces (e.g. safety measures, islanding, power quality etc.). Perhaps the biggest potential for reducing BOS costs lies in the possibility of siting photovoltaic power at the main points of electricity end-use by integrating it into domestic, commercial and industrial structures.

Another possibility is to use concentrators (lenses or other optical devices to focus sunlight on a small cell) to reduce the amount of expensive photovoltaic-active material. The photovoltaic community has long debated about the effectiveness of such systems; concentrators remain a viable option for significant electricity production only in regions with a high percentage of direct radiation, where it is possible to achieve a significant performance improvement.[4] The main problems related to concentrators remain the loss of diffuse sunlight (which cannot be concentrated), and the need for tracking systems and consequently for increased operation and maintenance expenses. The suitability of concentrators for desert regions is still an open question, but they seem unlikely to be of interest in European conditions.

---

[4] T. B. Johansson et al. (eds), *Renewable Energy Sources for Fuels and Electricity* (Washington DC: Island Press, 1993), Chapter 8.

Table 5.1 Long-run projections of PV costs in the United States

Year	Business-as-usual Capital costs (/kW)		Intensification of RTD Capital cost (/kW)		Cost at 1800 kWh/yr site (cents/kWh)		Cost, at 1200 kWh/yr site (cents/kWh)	
2000	3280	[2186]	2540	[1693]	–	–	–	–
2010	2290	[1526]	1770	[1179]	7.8	[5.2]	11.8	[7.9]
2020	1530	[1020]	1250	[833]	5.6	[3.7]	8.4	[5.6]
2030	1280	[853]	1010	[673]	4.5	[3.0]	6.7	[4.5]

Note: All costs, 1990 US$ and [1995 ecus] at 6% discount rate.
Source: Adapted from Idaho National Engineering Laboratory, et al., *Interlaboratory White Paper on Renewable Energy* (Washington DC: US Department of Energy, 1990).

Implications for PV costs

The number of promising avenues for cost reductions and efficiency improvements through development of better basic cell technologies, the scope for lowering BOS costs, and the cost savings expected simply as a result of expanding the scale of production, all mean that PV costs are expected to continue declining. In 1990, the US Department of Energy conducted an extensive appraisal of prospects for cost reductions, with the results shown in Table 5.1. This suggested that under a scenario of intensified RTD, the capital costs of installed systems could be reduced to below US$2000/kW (costs in ecus are given in the table) between 2000 and 2010 (US$1770 by 2010), with a reasonable prospect of being halved again by about 2030. The resulting cost of power, in mid-European insolation conditions of 1200 kWh/yr, could be about 11.8 ¢/kWh by 2010, and about 6.7 ¢/kWh by 2030 (all 1990 US dollars at 6% discount rate). At the best sites in southern Europe costs could be about 25% lower.

Williams and Terzian[5] later produced a scenario of accelerated development which achieved these goals somewhat earlier, with installed costs at 7.1 ¢/kWh (1990 US dollars at 6% discount rate, 4.7 ∈/kWh (1995 ecucents))

---

[5] R.H. Williams and A. Terzian, *A Benefit/Cost Analysis of Accelerated Development of Photovoltaic Technology* (Princeton University, 1993).

by 2020 for typical European conditions. The major European study *Photovoltaics in 2010* estimates current 'base case' costs at over 0.4 ecus/kWh (1992 money, 0.44 1995 ecus), and – focusing only on crystalline cell technology – sets a near-term target to reduce this to 0.25 ecus/kWh (0.27 1995 ecus) by 2005. Even allowing for the different units and cost bases, this is still substantially higher than the projections in the earlier US effort, which draws upon the less established but ultimately more promising thin-film technologies.

5.6 The potential market in Europe

From the discussion of economic and technological developments, it is apparent that competing with modern combined-cycle, coal, nuclear and even wind power solely on the basis of cost in a central station configuration is an unrealistic target for PV in Europe. Nor is the internalization of environmental costs, on its own, likely to change the picture fundamentally. As illustrated in Chapter 1, Section 1.2, clean-up of sulphur and NO_x does not add nearly enough to the costs of conventional generation to narrow the gap with PV costs significantly, nor would even the proposed European CO_2 tax or requirement for CO_2 removal. Thus, a successful strategy must capitalize on the unique characteristics of photovoltaics and the true value that these characteristics could produce for the utility and/or energy consumers. Over the coming decades, PV will only succeed commercially in markets where its special characteristics give it advantages over other electricity sources.

Off-grid applications

PV is already widely used for very small-scale applications[6] (microwave repeaters, cathodic protection, data acquisition systems, remote dwellings etc.) in which cost is not a major issue because other remote power options are limited and quite expensive, and the cost of extending and maintaining line power can be prohibitive. Figure 5.8 shows how the cost of grid supply rises

---

[6] G. Belli et al., 'ENEL's ten years' experience in the use of photo-voltaic generation', in *Proc. 11th E.C. PVSEC, Montreux* (Harwood, 1992); C. Jennings, 'PG&E cost effective PV installations', *21st IEEE PV Specialist Conference Florida* (Institution of Electrical and Electronic Engineers, 1990).

Figure 5.8 Cost of grid supply compared against PV for isolated dwellings

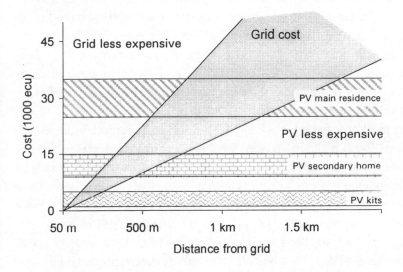

Note: This figure illustrates the situation in developing countries.
Source: ADEME.

with distance, and compares this with the cost of PV at different scales in a good solar regime. Main residences could generally require installations of at least 1 kWp. Applications in the power range 10 W to 1 kWp, in some cases up to 10 kWp, are already cost-effective today, and in most cases they pay for themselves in a six-month to two-year period. For technical and environmental reasons and as reliability increases and costs decrease, the size of this market will easily grow. In Italy there are already about 4000 remote customers supplied by PV stand-alone systems, and ENEL, the national utility, installed during 1993 (with support of the EC programmes VALOREN and THERMIE) 160 small stand-alone systems used to supply remote users at normal commercial conditions ('wireless electric service').[7] These applications are widely diffused in developing countries, where they

---

[7] G. Vaselli et al., 'Design and construction experiences', ISES Solar World Congress, Budapest, 1993.

represent the technology most appropriate to meet the energy demands of some rural settlements.

These initial markets for photovoltaics have been important in demonstrating their technological feasibility and environmental advantages, as well as in spawning a fledgling cell-manufacturing industry. There is still scope for considerable expansion even in Europe, where over a million people live in remote houses. With these, and occasionally occupied buildings (including summer residences), it is estimated that the European market for isolated dwellings amounts to about 150 MWp.[8]

Another market segment is for applications such as remote villages and small islands, with a typical power level ranging from 10 kW to 1 MW. These systems have a considerable potential in developing countries as well as in industrialized regions to supply minor islands, where the cost of energy and environmental considerations can in favourable conditions make such systems competitive. The EC's PV Solar Energy Pilot Project has already established a framework in which 15 plants (size 30 to 300 kWp with a total power of 1 MWp) have been built to supply electricity to small communities which are either remote or on small islands. No data were found on the potential size of such markets in Europe.

The markets for isolated systems are currently the most important, and will remain so for several years to come. However, the market for isolated systems in Europe will become saturated as prices fall, and will never represent a significant percentage of the total European electricity requirements. In the medium and longer term, PV must compete against grid-based power if it is to make major energy contributions. Given the higher per-unit cost of PV energy, the question is whether some kinds of grid-connected applications offer possible economic advantages compared with centralized generation.

Grid-connected applications: reinforcement and support options

One such market is the strengthing of the utility's distribution grid by use of photovoltaic systems.[9] Some pioneering efforts in this area have been

---

[8] European Commission, *Photovoltaics in 2010*, EC DG-XVII (Brussels: 1996).
[9] G. Rueger et al., 'Utility planning and operational implications of PV power systems', *Proc. IEA-ENEL Conference, Taormina, 1990* (Paris: IEA, 1992).

performed by Pacific Gas & Electric in the United States, for example with the construction of the 500 kWp Kerman PV plant to support the local sub-station (see Chapter 2). Systems for this market would have a power output from 100 kW to 1 MW, and they would be located at the electrical periphery of the utility system where they could support the local energy, capacity, voltage or reliability needs.

Because PV is modular and solar energy is widespread, this form of power is an ideal candidate for such dispersed grid-connected application at the 'edges' of the utility system. These systems are predominantly small and medium-scale generation units, characterized by physical dispersion of generation sites, according to the concept of 'embedded-generation' discussed in Chapter 2. In addition to electric energy generation, they may add substantial benefits ('grid-support') – reactive power support, reduction of transmission losses, enhancement of reliability, and delay in the need to upgrade the existing equipment. The more closely the grid load profile fits the daily and seasonal PV generation curve, the higher are such benefits, and the more the utility should be willing to pay per kilowatt-hour.

There are four distinct levels at which PV can be installed as distributed generation in a power system:

(1) connected to the medium-voltage (MV) distribution grid, near or within HV/MV substations (size: megawatts);
(2) connected to MV feeders (size: 100 kW to 1 MW);
(3) connected to the LV distribution grid, near MV/LV sub-stations (size: ~ 100 kW);
(4) connected to the LV distribution grid, at the customer site (size: a few kilowatts).

Institutionally, the largest systems are the easiest to manage. Several pioneering megawatt-scale photovoltaic plants are already generating reliable power to utility grids, including ENEL's 3-MWp plant at Serre (mentioned above) and a 1-MW plant in Vasto; and in Spain Union Fenosa, Endesa and RWE completed a 1–MW plant (financed by the EU and German BMFT) in Toledo. As with the early, multi-megawatt wind turbines, however, these plants are not direct forerunners of ultimately commercial systems. Rather, they provide field construction and operating experience and a realistic picture of current total-system costs and performance, and some financial flows for an industry that must first evolve through other routes and markets.

Systems for support at medium voltage or at MV/LV substations may often be located inside utility-owned sub-station areas, and/or in fringe areas. The distribution companies could analyse the options, plan the systems, and finance and manage their installations in sizeable contracts with suppliers. The value may be particularly high in the more remote and sparsely populated areas of southern Europe, with higher insolation, longer feed lines, and regional demand that may peak during sunlight hours (e.g. for air conditioning as well as industrial loads).

In general the economics of grid-connected PV will improve in accordance with two factors.

● The *lower the level* in the system — the closer to the point of end-use – the greater are the potential network-related savings arising from reduced transmission and transformation requirements, and hence the greater the value of the electricity.

● The *greater the integration* of PV with other structures and uses that offset the BOS costs, the less is the effective cost of the PV supply. Obvious examples include the installation of PV integrated with existing infrastructure such as building surfaces, motorway/rail embankments, sound barriers, car parks, recreational facilities, commercial and industrial installations etc., reducing or eliminating the cost of land-use and structural expenditure.

These two factors together point towards the fourth application (case (4) above) – scattered electricity generation from a variety of PV generators integrated with existing infrastructure at the customer site – as offering the potential for exceptional economic advantages, if the many complexities can be overcome. These benefits could be supplemented further in applications which require DC power at appropriate voltages, such as much electronic equipment. To the extent that this could avoid completely the need for transformers (in the end-use equipment) and inverters (in the PV system), this would yield both capital savings and reduce losses in both the supply (PV) and end-use equipment – a more efficient system all round. As noted in Chapter 2, local DC systems are a technical possibility; AC systems only evolved because of the economies of scale arising from centralized power generation.

DC appliances are available, including ones developed for remote applications preferred by PV.[10] They are generally more energy-efficient then conventional appliances, but also very expensive, presumably because of the much smaller market. In general, employment of PV for many end-uses may be associated with more energy-efficient equipment, so that the overall value of the contribution would be greater than the electricity delivered would imply. Irrespective of the complexities associated with direct DC applications, the potential benefits of siting at end-use are readily apparent. It is for this reason that siting on buildings has begun to attract particular attention, as we now consider in more detail.

Building applications

Domestic and commercial buildings are increasingly becoming major users of daytime electric energy for lighting, appliances, office facilities (computers etc.) and air conditioning. PV can be placed on existing buildings (retrofit), for instance at the same time as maintenance work is being done, or integrated into new buildings from the design stage. These applications are low-voltage grid-connected systems of 1–50 kW per site for residential and commercial building. They may be particularly attractive if they can be structurally, architecturally and electrically integrated into buildings and their surfaces (terraces, rooftops, façades, sky-lights etc.). At the point of end-use, the value of such generation is very close to the marginal cost of supply, which in principle should be close to the sale price of electricity. As photovoltaic costs decline, this application may also become an attractive opportunity as a measure for demand-side management.[11]

There are two main variants of building-mounted PV. One is PV façades, or 'PV cladding', in which PV is integrated into the south-facing building structures. Savings in the costs of alternative claddings are added to the

---

[10] *Photovoltaics in 2010* (Note 8).

[11] A. Sorokin et al., 'Assessment of PV rooftop integration into southern Italian architecture', CEC Conference on Solar Energy in Architecture, Florence, 1993.

Figure 5.9 Net electricity costs of PV cladding

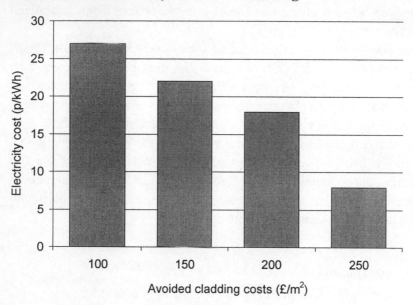

Source: Robert Hill, personal communication.

benefits of on-site generation. Figure 5.9 illustrates potential net electricity costs from PV systems when various levels of avoided cladding costs are credited to the PV system. This appears particularly suitable for commercial and service-sector buildings in central and northern Europe, which often have daytime electricity loads. One factor which could limit PV applications in this sector is that larger buildings are increasingly also turning to gas-fired co-generation of heat and power as a cheap option for on-site energy. On the other hand, even in these cases, one of the difficulties is that the electricity output is tied to heat demand, which is heavily concentrated in winter (and early mornings). Ironically, therefore, such loads may be one of the few end-use applications in Europe in which PV may have positive daily and seasonal correlations with the net on-site demand – possibilities which would require further and site-specific investigations.

The other main variant is rooftop mounting. This may also come in a number of variants, from simple retrofits to PV tiles. The most extensive

Table 5.2 Solar building in European countries: surface areas and technical PV resource

Countries	Irradiation (k/Wh/m² per year)	Net available rooftop solar surfaces			Installable PV capacity[a] (GWp)	Potential PV electricity[a] (TWh/yr)	PV rooftop potential electricity[a] (%)
		Houses (km²)	Offices services (km²)	Industrial buildings (km²)			
Austria	1200	50	15	13	9.6	8.1	18.9
Belgium	1000	43	20	14	9.6	6.7	11.5
Denmark	1000	34	11	6	6.3	44.1	15.8
Finland	900	45	11	8	8.0	5.0	7.8
France	1200	362	122	85	70.3	59.0	18.3
Germany	1000	532	214	242	121.6	8.5	15.5
Greece	1500	64	11	6	9.9	10.4	29.8
Iceland	800	2	1	0	0.4	201	4.5
Ireland	1000	16	5	4	3.1	2.2	15.1
Italy	1300	336	120	86	67.6	61.5	24.5
Luxembourg	1000	2	1	1	0.5	339	7.5
Netherlands	1000	63	30	21	14.0	9.8	12.2
Norway	900	34	10	8	6.4	4.0	4.5
Portugal	1700	54	11	11	9.6	11.4	42.4
Spain	1600	145	60	51	32.0	35.7	24.2
Sweden	900	78	20	13	14.0	8.6	6.6
Switzerland	1200	42	18	12	8.8	7.4	23.8
UK	1000	248	123	96	58.0	40.6	12.3
Europe		**2150**	**803**	**677**	**450.0**	**360.0**	**16.3**

[a] Based on 1990 electricity consumption.
Source: EC, *Photovoltaics in 2010*, derived from Tables 2.23, 2.25, and 2.26 (1996).

and promising applications may be in southern Europe, where there are more flat roofs and higher sunlight inclination, and may include housing.

The theoretical potential for such systems is very large. Table 5.2 shows the estimated available 'solar surface' in European countries, along with an estimate of the theoretical potential PV generation that this represents. In the Mediterranean countries, this technical 'building resource' is in the range 25–40% of current electricity demand. In most of central and northern Europe the resource is in the range 12–18%, declining below 10% in Scandinavia. Averaged across western Europe it is 16.3% of 1990 electricity consumption. In general, houses account for a little over half the total, and offices and services around 20% of the total building surface.

These figures are of course highly stylized, and there would be many constraints on the practical contribution. The resource would increase with the still-expanding building stock, and with improving cell and system efficiencies. Specifically, the resources indicated in the table are not constrained by the electricity demand of separate buildings. Clearly if PV systems were exporting power to the grid, many of the benefits of embedded generation would be lost, though not all if the power were consumed elsewhere in the local system. Nevertheless, this must set considerable limits on the realistic resource, particularly for rooftop systems on domestic dwellings that may have peak demand in the evenings. One study also suggests that PV in buildings will be most attractive if combined with a limited amount of on-site storage.[12]

To foster the wide adoption of these systems, a few aspects of integration need further investigation.

● Architectural and bio-climatic integration of PV modules in buildings: this may require the lifetime of PV components to become comparable with that of other standard structural elements (i.e. about 30 years). The ultimate goal may be represented by the so-called 'PV-tile'. After some experience with amorphous silicon integrated roofs, most studies are aimed at developing the practical technological aspects of water-tightness and easy electric connection.[13]

● Effect of orientation: except where the building and roof inclination happen to be ideal for solar generation, output may be increased by orienting the modules specially to capture maximum sunlight; it remains unclear whether and when this may be economically justified.[14]

● Façade integration: an integral façade can give extra value to the electric power production by allowing summer shading with overheating

---

[12] John Byrne et al., 'Evaluating the economics of PV in a demand-side management role', *Energy Policy*, Vol. 24, No. 2 (February 1996).

[13] C. Meier et al., 'Solar tile: a special PV module integrated in clay tile roofs', *Proc. 11th PVSEC, Montreux, 1992* (Harwood); C. Meier, 'Plug-socket-interconnection system for PV-modules', *Proc. 12th PVSEC, Amsterdam, 1994* (Bedford, UK: H.S. Stephens and Associates).

[14] G. Chimento et al., 'PV roof for a laboratory building', 3rd European Conference on Architecture, Florence, 1993.

protection, but it requires careful design to obtain pleasing effects.[15]

● Safety aspects: these cannot be underestimated, given that scattered PV power production occurs close to generally unskilled users, with the related electrical hazards and risks of fire. The International Electrotechnical Commission has prepared a code, 'Safety Guidelines for Grid-connected Photovoltaic Systems Mounted on Buildings', but a continuous effort is needed to overcome national differences which, at the moment, do not allow a general application of the safety rules required for PV generation.

Only a growing confidence developed through the widespread diffusion of the technology will give architects and others the possibility of choosing photovoltaic roofs/façades as viable options in their new designs.

Transport-related applications

As noted, buildings are a special case of the general principle that the economics of PV improve insofar as they can be integrated with existing infrastructure and applied directly to end-uses. In Europe, many other infrastructure requirements create significant areas that could readily incorporate PV at no extra balance-of-system cost. Examples could include motorway verges and railway embankments. Indeed, these often require sound or other kinds of barriers, which could readily incorporate PV. Several schemes incorporating PV into motorway barriers have been completed.[16] Dutch calculations suggest that about 18 TWh – some 20% of Dutch electricity demand – could in theory be generated from road and railway verges in the Netherlands.[17]

It is interesting that these primary examples of integrated siting are transport-related. At present, some railway systems in Europe are electrified (requiring high-voltage DC) and roads require lighting (obviously, not correlated with solar input!) but hardly any driving uses electricity.

---

[15] M. Pornansky et al., 'Building integrated PV systems: examples of realized PV-roof and PV-façade power plants with specially conceived PV-modules for building integration', *Proc. 11th PVSEC, Montreux, 1992* (Harwood).

[16] These include the 100 kWp project on the N13 motorway near Chur in Switzerland, a 40 kWp installation near the A1 in Austria, and a 55 kWp installation in the Netherlands. *Photovoltaics in 2010*, op. cit. (Note 8), p. 20.

[17] *Photovoltaics in 2010* (Note 8).

Nevertheless, concerns about various transport-related problems, including noise and air pollution, have led to an interest in electric or other non-gasoline vehicles. The electrification of railways also continues apace.

Thus a possible emerging end-use application could be for powering electric transport. The practicalities of using PV for electric public transport deserve further investigation. A bigger application could possibly be the charging of vehicle batteries. For this, PV could be mounted on the façades and roofs of car parks (also providing shade and shelter in otherwise open-air parks, and inside for lighting), and along roadside verges to supply charging points at service stations. A great attraction is that these are natural DC applications, with savings on transformer and inverter costs and losses. On-site electrolysis of water to produce hydrogen could be an alternative if hydrogen-powered fuel-cell vehicles were preferred (see Volume III). The size of the market would be inextricably linked to the market for non-petroleum vehicles, which is very small at present. The use of PV-supported vehicles might be stimulated by air-quality policy, or by the approaching spectre of concentration and depletion of oil resources, but it will not appear spontaneously.

A third transport area could be water transport, particularly marine pleasure boats and perhaps some inland barges, where there is a demand for power that is mobile – and above all (for inland waterways) clean and quiet. Unlike in the case of road vehicles, significant energy might be collected from the area of the boat itself, though some quayside constructions could possibly also be used to collect power for battery-charging points. The market is small compared with other transport applications, but has a high value.

These possible synergies between such siting and transport-related DC markets are speculative, and no serious market analysis appears to exist. Both railway and road vehicle applications would clearly depend heavily upon complex developments in transport technology, policy and infrastructure. Transport markets, being predominantly non-electric, are explored further in Volume III, where the discussion concentrates mostly upon biofuels. Given the manifold debates on transport, the relationship between PV and transport would appear deserving of further study.

The scale of potential markets for PV: conclusions

The economic dependence of PV upon favourable market applications, and on further extensive technological and systems development, makes it impossible to project a meaningful estimate of the long-run potential. At the minimum, the conditions summarized above suggest that foreseeable developments in technologies and markets will enable PV to be competitive, without major increases in energy prices, for wide-ranging remote applications, for some grid reinforcement in southern Europe, and for some level of contribution to building electricity demand, particularly for service-sector buildings. Together these applications could plausibly amount to a few per cent of European electricity demand.

An upper limit is harder to define. It seems implausible that PV will ever be a sensible way of meeting needs for intrinsically concentrated industrial electricity demands, which would require collection of solar energy from widely dispersed sites on a crowded continent, for transmission in competition with a host of other sources that by nature are more suited to centralized production. Obviously, PV is an inappropriate form in which to supply dispersed heating needs, for which direct use of solar heat makes far more sense.[18] Yet probably as much as half of European energy demand is associated with dispersed uses of high-grade energy – for lighting and other energy supplies to buildings, and for transport applications where, as outlined, there are also interesting synergies between siting opportunities and possible DC applications. In principle, PV could have advantages over competitors for such applications. The gap between the few per cent of electricity that could be derived from foreseeable PV applications and the full range of dispersed energy demands for which PV could be theoretically available and appropriate can only be narrowed by as yet unforeseen innovations in technology and systems, and by time and experience.

Finally, it must be remembered that the European market pales into insignificance beside the potential for PV in the rest of the world, particularly in developing countries. Quite apart from grid-connected applications there is

---

[18] Note that solar heating need not in principle conflict seriously with PV collection. Even a 15% PV collection efficiency would leave the great majority of solar energy still available for heating applications.

a huge off-grid potential: a billion people are still not connected to grids. The global picture is discussed below, in Section 5.8. First, we consider the industry and policies in Europe as they stand today.

5.7 European photovoltaic industry structure, programmes and prospects

The European photovoltaic industry's contribution to total world module shipments grew steadily in the 1980s, reaching 20% in 1995. More than 80% of the products come from six major manufacturers (Table 5.3). These are part of a group of 17 in the world that each shipped more than 1 MWp in 1995, and compete in the world market. Manufacturing technology is mostly based on crystalline silicon and focuses on modules for power application. Very little amorphous silicon and very few related devices are produced in Europe.

The European photovoltaic industry is leading in several technological sectors. Photowatt pioneered expertise in wire-sawing, BP-Solar established a pilot line for the production of high-efficiency laser-grooved buried-contact cells, and Eurosolare has led in production of high-quality polycrystalline ingots. The financial capability of the photovoltaic industry is strongly based on corporate affiliation, often with oil companies (Eurosolare/AGIP, Photowatt/Shell, BP Solar/BP). However, for most of the 1990s these companies have been operating at a loss, and this puts them under continual pressure from the parent company. Current difficulties need to be overcome without losing the know-how acquired.

There is a great need for cooperative action in the industry, with a more coordinated effort to attain common goals (an approach being pursued, to some extent, in the US and Japanese national photovoltaic plans). Since the present industrial technology process is long and complex, and requires various types of expertise, it is not easy for a single company to be effective both in RTD and in manufacturing throughout the production cycle. So one approach is to explore the ideas of pooling resources and segmenting the production process. These concepts could be implemented through a network of industrial companies, each operating in a specific sector of the production cycle, but all acting with a single set of guidelines and aimed towards the same programmes and objectives.

Table 5.3 PV modules shipments by major European manufacturers, 1995

	Europe (kWp)	Percentage of world total
BP Solare Espana	3,250	4.5
Eurosolare	2,700	3.8
Photowatt	1,850	2.6
DASA/TST	1,800	2.5
Isototon	1,500	2.1
Helios	400	0.6
Siemens Solaer GmbH	200	0.3
Other	2,900	4.0
Europe	*14,600*	*19.4*
World	72,000	100.0

Source: ENEL.

The other strategy should be to address the problem of finding a cheaper and more effective technology, which could ultimately open larger markets. The solution of this problem is not yet clear; many options are still possible. Again each single company has only limited experience and opportunities to find the right solution at the right time. The support of a pilot industrial line to demonstrate the new technology is perhaps the key step. Of course such proposals are feasible only in the framework of strong political initiative aimed towards the growth and success of the European photovoltaic industry.

PV electricity generation in Europe in 1994 amounted to about 0.05 TWh/yr (installed capacity: 67 MWp), which corresponded to about 0.003% of the annual European electricity consumption, and was enough to supply about 50,000 people. The manufacturing capacity for PV generators in Europe is at least 35 MWp/yr, or about twice the current annual market needs. For PV technologies to make a significant contribution to the energy economy of the European Union, key European, national and regional agents must stimulate the market for PV generators, the level of PV component manufacturing must be greatly expanded, and the PV skill base must be strengthened.

This situation – the long-term hopes for the technology set against the difficult market conditions and extensive existing investment in manufacturing capacity – has led to a variety of national programmes for supporting

Box 5.2 The Swiss photovoltaic programme

In 1991 the Swiss government launched a national programme with ambitious objectives for renewable energy which established targets for new renewable energy sources in the year 2000 to supply 3% of the heat consumption and 0.5% of electricity generation in Switzerland.

It is planned to cover about one quarter of this 0.5% of electrical power by PV power systems. This would require 50 megawatts (peak output) of PV capacity, producing about 50 GWh annually.

By the end of the year 1994 the existing capacity of PV grid-connected plants in Switzerland reached 4.8 MWp. The annual PV electricity production was 3 GWh. To fulfil the demanding goal of 50 MWp for the year 2000 according to the national programme ENERGY 2000, many supplementary measures had and will have to be taken. The Swiss federal 'Promotion programme for PV' includes the following activities.

Pilot and demonstration projects. A substantial number of impressive pilot and demonstration projects including new approaches of building integration and infrastructure have been planned and built in Switzerland.

Improvement of the basic legal conditions. The energy utilization resolution passed by the Swiss parliament on December 1990 also creates new, improved, basic legal conditions for the use of photovoltaic technology. The public power supply companies are obliged to purchase the electrical energy produced by self-sufficient users and individual producers using photovoltaic, wind, combined heat and power, and mini-hydroelectric power stations and to reimburse them at an appropriate rate.

An initiative has been launched to establish a national policy for the cost equalization for all new grid-connected PV installations. This approach should drive the market forces from investment subsidies to payments for the actual energy produced (~ SFr0.2/kWh – comparable to 0.13 ecus).

Information and training. Within the context of the national programme of action for building and energy, the PACER incentive programme is concentrating on renewable energy sources. The PACER programme, in close cooperation with commerce and industry, schools and the federal government, promotes the uses of renewable energy sources and is intended to provide engineers, architects and installation engineers with the knowledge needed to apply photovoltaic technology. To transfer know-how and training to vocational schools, a special programme has been launched to promote PV plants financially.

PV school buildings. The federal programme allows a financial support of SFr5,000/kWp (comparable to 3,235 ecus) to promote grid-connected PV plants for school buildings. By the end of 1993, 43 individual projects with a total capacity of 250 kWp had been approved; 16 were already connected to the grid and seven were under construction; 14 were planned with a capacity of 100 kWp. Gearing towards this group of buildings is a strategic choice. Schools are normally owned by the local authorities. This means that the benefits of using photovoltaic technology are brought to the attention of the local authorities, teaching staff and students.

Source: Derived from 'Increasing the contribution of PV in Switzerland', *CADDET Newsletter* (November 1995).

Table 5.4 Targets set for PV in Europe

Parameter	PV2010	ALTENER	TERES		Madrid Plan
			Existing	Proposed	
Year	*1994*	*2010*	*2005*	*2010*	*2010*	*2010*
Installed PV in Europe (MWp)	70	2000	500	1135	6730	16,000
PV electricity (TWh/yr)	<0.05	2.0	1.0	1.25	7.4	
Manufacturing capacity (MW/yr)	35	1000	–	–	–	2000
European share of international trade in PV	>30%	>40%	–	–	–	500 MWp

Note: All targets apply to EU-12 except the PV 2010 study, which includes the three countries that joined in 1995
Source: European Commission, *Photovoltaics in 2010*, Table 7.

PV, and projections. The biggest national programme in the EU, the German '1500 roofs' programme, had by 1995 achieved the target its name implies, with PV systems (size from 1–5 kWp) connected to one-family houses and small company premises. The project was funded by the BMFT (RTD ministry), but its cost makes it unlikely to be extended in the current financial climate.

This and many other national programmes are summarized in the Commission–industry collaborative study *Photovoltaics in 2010*. One of the most varied and thoughtful of these, widely hailed as a model national programme, is the Swiss one summarized in Box 5.2. It aims to demonstrate varied applications, including a focus upon schools, to raise widespread awareness of the technology and its possibilities. Nevertheless, progress since its inception, with 3 MW installed by the end of 1994 towards a goal of 50 MW by 2000, does illustrate the constraints and the small scale of PV activities so far.

These realities help to shape the medium-term goals established for European PV capacity, summarized in Table 5.4. The ALTENER programme establishes a goal of 500 MW by 2005, while the TERES study projects a capacity ranging from 1,135 MW under existing programmes to 6,730 MW under the full set of 'proposed policies', with corresponding generation amounting to 0.1–0.4% of the 1990 European electricity supply.[19] The Madrid

---

[19] See Volume I, Chapter 3, for a description of these studies and programmes.

Figure 5.10 The diffusion strategy of PV systems application

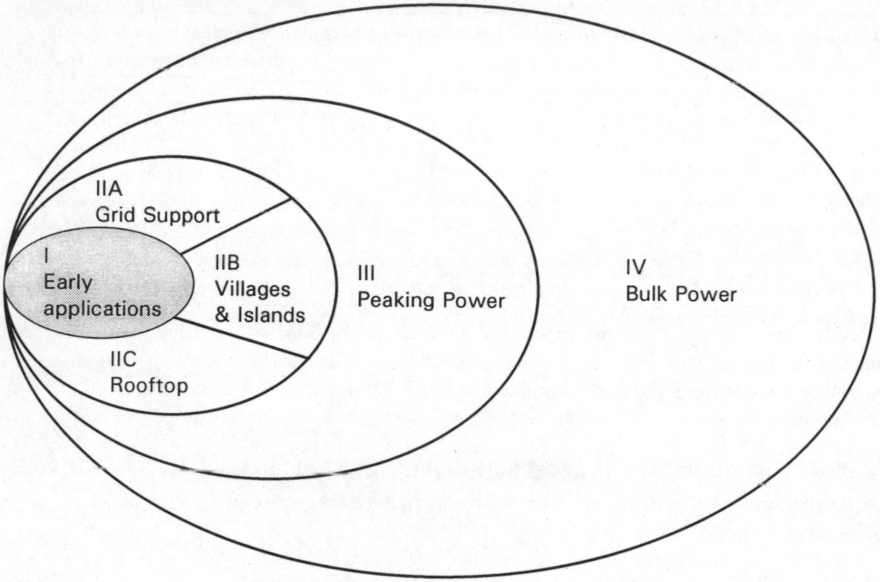

IIA
Grid Support

I
Early
applications

IIB
Villages
& Islands

III
Peaking Power

IV
Bulk Power

IIC
Rooftop

Source: J. Iannucci et al., 'Potential of PV systems for present and future electric utility applications', IEA conference, 1990, Taormina, Italy.

Declaration calls for 16,000 MW by 2010, which would represent 1% of supply and expansion at about 45% per year from 1995 levels. *Photovoltaics in 2010* suggests a much more modest goal to reach 2000 MW by that date, an annual growth rate of 20%, with a suggested manufacturing capacity by then of 1000 MW/yr requiring investment of 2,000 million ecus.

In tandem with the PV development of technology and industry, thinking about PV promotional strategies has evolved. Efforts at first focused mainly upon reducing cell costs, with an implicit assumption that the main grid-connected application would be for power production along conventional lines: a PV 'power station'. As cell costs fell – but not as quickly or as easily as hoped – the 'balance-of-system' costs became more visible and the economic difficulties became more apparent. Greater awareness of the different potential markets and siting possibilities led to more refined conceptions of PV development strategies. In 1990 a conference of the leading analysts set

Box 5.3 The diffusion model of PV market development

Greater awareness of niche markets has led to growing acceptance that PV must progress through a series of expanding market segments – including small- and medium-sized, and both remote and grid-connected applications. Expanding markets are seen as the key to achieving the required reduction of costs, for which a larger manufacturing capacity and continued RTD will be required. In 1990, at the IEA Executive Conference on 'Photovoltaic Systems for Electric Utility Applications' held in Taormina, Italy, [a] the participants endorsed a five-part parallel diffusion strategy (Figure 5.10) which identifies the various applications most suitable for power generation (referred to as market segments or 'niche' markets) and describes the ways various factors will affect the rate of their deployment.

Beyond the early and small-scale applications (such as consumer electronics and communications), this strategy envisages diffusion through five main market segments:

● Isolated applications, such as remote houses and small islands or mountain villages, in which the costs of PV are offset against the much higher costs of grid extension.
● Segments in which photovoltaic systems compete on the demand side of the meter, particularly buildings.
● Grid support, embedded at strategic points within electricity distribution systems.
● Peak power supplies in regions with solar-correlated peak demands.
● A long-term goal is the aspiration for PV to compete in the form of power plants for centralized grid supplies. As discussed earlier in this chapter, this seems an unlikely goal in Europe but could be feasible in tropical or desert conditions.

All five market segments are, to some extent, being developed and explored. It is debatable whether or when the last two in particular can ever be economically feasible in European conditions. Nevertheless, the diffusion model can be considered a suitable short/medium-term guideline that incorporates some important attributes: market-driven technology development, steady market growth, growth of the PV industry infrastructure, a broad and successful utility experience base with PV systems, and strong corporate acceptance.

[a] Proceedings of the IEA-ENEL Conference on PV Systems for Electric Utility Applications, Taormina, 1990, op. cit. (Note 9).

out the 'diffusion model' of PV development, based upon progressive colonization of niche markets (Figure 5.10 and Box 5.3).

Informed by this perspective, but driven also by the difficulties the industry has been facing, the 1990s have seen various 'action plans' proposed for promoting PV developments in Europe. The two main ones have been the PV chapter in the Madrid Action Plan,[20] and the strategy set out in

---

[20] EU Commission, *An Action Plan for Renewable Energy Sources in Europe* (Madrid: DGXII-DGXVII, 1994). Also see European Solar Council, *The Club de Paris: A European Action Plan for Solar Energy* (Paris: 1993), which helped to lay the basis for the Madrid plan.

Photovoltaics for 2010. These differ widely in their ambition and detail, but contain many similar elements.

(1) *Long-term commitment and target for PV generation in Europe*: A long-term political commitment to target levels of grid-connected PV electricity generation in Europe (see Table 5.4 for various targets) is called for, to create a positive climate for investment in advanced low-cost/high-volume PV manufacturing facilities in Europe.

(2) *Ten-year transitional EU support package*: The political commitment must be complemented by a ten-year 'transitional package' of publicly supported programmes to facilitate the move to more competitive grid-connected PV electricity generation through the scaling-up of European PV module and system component manufacturing. The package should involve enhanced collaboration between the EC and member states and funding, to support a substantially expanded combination of the previous CEC JOULE, THERMIE, VALOREN and Aid programmes. Elements would include:

● continuing support for RTD;
● enhanced programme of demonstration and technology promotion projects in the EU;
● international dissemination projects and technology promotion;
● capital subsidies for PV rural electrification in sunny regions of the EU; and
● capital subsidies for PV exports to developing countries.

(3) *EU–industry partnership to finance new low-cost PV manufacturing plants*: To realize cost reductions from a scaling-up of manufacturing capacity, the strategies call for a new initiative to promote rapid investment in new PV module manufacturing plants in Europe, through an EC-led tender for a jointly funded 'partnership investment' programme for new PV manufacturing plant. The provision of the financing would be linked to firm contracts between the industry proposers and European utilities to supply PV power generation in the EU, thus encouraging long-term collaboration between the utilities and the European PV manufacturing industry.

(4) *Financing of PV power generation in the EU*: This would address stimulation of the demand side. The strategies suggest encouragement of utility cross-subsidization for utility PV investments – for example that the utilities should be permitted (with EU-wide agreement and co-ordination) to charge an ad-hoc PV levy and to make a small management charge for their PV support activities. Also, they call for special tariffs for independent PV generators.

In general, this question of demand-side stimulation is the least developed aspect of the strategies. The *Photovoltaics in 2010* study does, however, address more closely the focus of markets. It advocates a subsidiary target of 500 MW of grid-connected PV to be installed by 2005, divided in specified ways between façades, rooftop mounting and grid-support applications, with a public capital subsidy initially 49% of installation costs in 1996, down to 30% by 2000 and then declining linearly to zero after 2006. The total subsidy required over the ten-year period is estimated as about 500 million ecus. Not surprisingly, such a programme has yet to be approved.

The European situation can be compared with efforts in the United States and Japan. Both of these countries have made a long-standing commitment to promoting PV RTD. The industry in the United States is considerably larger than in Europe (though a significant part of it is focused upon amorphous silicon, so far without the success promised), and continues to enjoy far greater governmental support than in Europe. In Japan, MITI is funding a programme of 70,000 roof installations, with capital subsidies of 49% available for households which take up the offer.

Thus, unlike wind energy, European industry in general does not enjoy the most favourable conditions or have a clear competitive advantage. Nevertheless, many technical records are held by European companies, and despite greater support elsewhere Europe has kept a creditable share of 20–30% in the global market, which the *Photovoltaics in 2010* strategy aims to expand to 40% by 2010.

5.8 Global perspectives

These European goals and concerns are, ultimately, dwarfed by the global picture. The great majority of solar energy resources, of villages without power, and of projected power system expansion lie in the developing world. It is in these regions that PV could make the greatest contribution to development aspirations and to global environmental goals; and the developing-world markets from this perspective would dominate global sales. By the same token, these markets would be key to the large-scale production that will to bring down costs.

This has led to many ambitious proposals for meeting the basic energy needs of the world's poor by means of PV. A major UN-sponsored workshop in São Paulo in 1991 unanimously agreed that PV crystalline silicon technology is 'mature enough to meet the needs of inhabitants in remote areas of both industrialized and developing countries'.[21] They formulated the goal to be met in a joint solar programme: the dissemination of one million PV systems (reaching 1% of the world's rural people) by 1995 and, by the end of the century, reaching 10% of rural poor. The rationales cited included the goals of sustainable development, abatement of climate change, and international security. The European Commission has also promoted a 'Power for the World' vision of providing PV to a 'million villages' that are currently without power. These schemes call for international funds to be established in pursuit of these goals.

Long-term and more ambitious world-wide targets are foreseen in some recent studies suggesting a way for photovoltaics to become an energy-significant option for the future.[22] The key point for the success of such a strategy is again to bring down photovoltaic prices quickly via a large-scale market development and a strong and accelerated RTD effort. Required funding was estimated at around 4–6 billion ecus for the EC system. The total installed capacity foreseen in this accelerated development scenario could be around 400 GW by 2025, more than half of which is in industrialized countries, with utilities becoming the biggest purchaser in the photovoltaic industry.

---

[21] International Workshop on Mass Production of Photo-voltaic – Commercialization and Policy Options, held in São Paulo, Brazil, September 1991. The workshop focused on the identification of measures to foster PV rural electrification in developing countries. This workshop brought together 141 participants from 22 countries, both industrialized and developing, and seven agencies of the United Nations. Published as R. Hill, I. Chamboulyvon and E. Omeljanovsky (eds), 'Prospects for Photo-voltaics – Commercialization, Mass Production and Application for Development', in *ATAS Bulletin* 8 (Geneva: UN, 1992).

[22] Williams and Terzian, op. cit. (Note 5).

As yet, however, progress has been very disappointing for PV supporters. Rather than mushrooming, global PV shipments almost stagnated during the early 1990s, although growth has now resumed. Despite the potential 'peace dividend', the visions of a new world order in the 1990s and the Rio Earth Summit, the first goal of the São Paulo workshop has passed with little change; its aspirations for 2000 are patently and grossly unrealistic. One directly relevant government-sponsored international fund – the Global Environment Facility – has been established, and it has pursued some PV projects, but on a scale that is negligible compared with the global ambitions, and the progress even of these projects has been criticized.

One core political problem is that the promoters of global PV development for rural electrification have as yet failed to establish a significant political constituency in the rich countries. Given the costs, the development charities remain far from convinced that it would be a good route to poverty alleviation; and governments have displayed no interest in spending taxpayers' money on such grand global schemes when the taxpayers themselves express neither understanding of it nor interest. Calls for a grand global development of PV for development fall foul of the realities of electoral and bureaucratic politics in the developed countries.

There remain as well the problems of implementation. The litany of difficulties associated with renewable energy expansion in the developing world has been sketched in Volume I, Chapter 5. Bluntly, in most cases developing countries have other priorities. It is not easy to change the structure of investments from the outside, even given finance and political will in the donor community, for the reasons set out in Volume I, Chapter 5. Where durable success has been achieved, it has generally involved a high degree of local promotion and involvement, aided but not driven by international donors. The relative success of PV in Sri Lanka, sketched as a case study in Volume I, Chapter 7, is an example of this; supply to remote villages in Indonesia is widely considered as another success story, and others are beginning to emerge. PV is thus starting to take root as a reliable, credible option for supplying power to remote homes and villages. But funding remains a major constraint, and the scale and pace of these developments fall far short of the global targets and visions promoted by the PV community.

Williams and Terzian take a rather different tack towards justifying global strategies. They argue that schemes for sponsoring rapid global development of PV can be justified not only in terms of rural development, poverty alleviation and sustainability.[23] They analyse a strategy of PV development via the steady colonization of 'niche markets', and compare it with a more ambitious strategy of 'accelerated development' which involves far greater sponsorship of PV applications together with an intensified RTD effort. By estimating the rate of cost reductions in each case, they conclude that the strategy of accelerated development is ultimately justified by the economic returns from the more rapid development of the technology.

This is an important conclusion and there are signs that this analysis has had an influence on the US strategy, particularly in terms of domestic RTD support, and on the Washington-based global financial institutions. In 1996 the UN/World Bank Global Environment Facility together with the International Finance Corporation announced the 'PV market transformation' initiative, establishing a fund of about $30 million targeted for supporting PV applications primarily in Morocco, Kenya and India. In view of the past difficulties with supplier-driven 'technology transfer' projects (as summarized in Volume I, Chapters 5–7), the money is to be made available to winning bidders among indigenous companies that respond to the call for tender, with a credible marketing plan for large-scale PV applications that can bring additional indigenous and international resources to bear. At the time of writing, the first tenders are awaited.

This is probably the most important initiative to date. However, larger and longer-term expansion cannot depend solely upon targeted support from the official global development and financial institutions, whose main criteria are poverty alleviation and support for least-cost development investments. More general proposals for grander global PV strategies face the fundamental problem that there is no coherent global industry and global government to support it. Investment in PV is ultimately justified by its global prospects, as it is the renewable energy with the greatest global potential of all. But in the next century, global visions will only turn into reality through the efforts and promotions of individual, profit-seeking companies together with specific governmental institutions. Which brings us back to the role of Europe.

---

[23] Ibid.

5.9 PV in Europe: a strategic view[24]

The report *Photovoltaics in 2010* represents the most extensive and comprehensive analysis of PV technology, industry and potential markets yet published in Europe. Both this and the Madrid Action Plan set out proposed plans for promoting PV developments to the year 2010. Although they differ greatly in the level of ambition, these plans share many common elements, as listed above. Rather than replicating such an effort, this section tries to place PV technologies and strategies in a broader context, highlighting issues that still need resolution and providing a longer-term perspective.

Despite far more extensive and longer-term public and private RTD effort, the development of PV as a major energy source in many respects lags well behind that of wind energy. The goal set in the study *Photovoltaics in 2010* equates to one large power plant; it is dwarfed by the capacity of small hydro (and biomass and waste) plants, and is on a par with the wind energy capacity installed by 1994. PV remains more costly than most other renewables, European resources are poor by international standards and the European PV industries have few inherent competitive advantages.

Given these considerations, will PV really make a significant contribution to future supplies, and will governmental efforts to support PV really yield worthwhile returns? The sheer scale of the energy resource, the growing global market, the steady progress of the technologies and the many and varied market applications suggest good prospects. But this conclusion should be qualified by recognition that, to be successful, European strategies need to be marked by their sophistication and relationship to credible long-term markets. For PV is not only the most expensive of the onshore renewable energy sources. It is also the one which differs most from any other kind of power source.

As argued in Chapter 1, the future institutional context for PV in Europe – and probably in much of the rest of the world – will be set by the progressive liberalization of electricity systems. In this context, individual utilities are no longer likely to provide extensive strategic RTD of interesting technologies; governments will need to support important, nascent technologies

---

[24] This and the subsequent sections were written by Michael Grubb and the opinions expressed should not necessarily be attributed to the other author of this chapter.

in various ways, if they are to progress competitively against established technologies. Yet there will be clear political limits to how much expenditure, explicit or as hidden subsidies, governments will bear. At the same time, there could be far greater scope for non-utility agents to invest and innovate, if they can generate adequate financial returns; and the evidence of wind energy is that, given the right financial conditions, this route could yield far more rapid expansion than expected. But to achieve adequate returns, PV will need to utilize the best technology and extract the full added value from its differing characteristics.

This will ultimately require changes in systems and institutions far greater than is generally realized. A self-sustaining market for PV – one that is not dependent upon the continued graces of demonstration and diffusion programmes and special PV tariffs – will only emerge if and as the concept of 'distributed resource' systems sketched in Chapter 2 emerges as a reality. The engineering components are now mostly developed, but they need demonstration and implementation as integrated systems, including the possible role of PV. It will take many years to gain engineering and user familiarity with the best ways of doing things. Innovative distribution companies will need to be at the forefront of such developments.

Yet the engineering issues are probably modest set against the institutional ones. For whereas PV used for grid support can be planned, financed and managed by the distribution companies, PV for grid-connected end-uses can only emerge as a self-sustaining market if and as liberalization extends down to the level of those end-uses. This could enable service-oriented companies to appear and provide systems which could include PV to meet customer needs. This could start most readily at the level of offices and other service-sector buildings, such as schools and hospitals. It could in principle proceed to dwellings.

Furthermore, these potential users would need to gain not only economic advantage but also familiarity and confidence. Users – whether distribution companies buying for grid support or building designers, owners and operators, or whatever – would have to be able to purchase and install photovoltaic systems with the same certainty that they now have in purchasing and installing conventional equipment. The ALTENER programme has established procedures for contract-based 'guarantee of solar results', which could help in the process of increasing confidence in the hardware, the suppliers and the operation of the equipment, built up over years of successful design, testing and operation.[25] The changes required for transport applications – involving a shifting of fuel and vehicle technology for private transport – are even greater. Even if these applications to private transport can happen on a large scale only when replacements for petroleum fuels are required, the lead times could be such that exploration of the market possibilities might need to begin quite soon.

In the long run, the emergence of local DC systems, in which PV would really come into its own element, is likely to require further institutional changes, including the involvement of the manufacturers of end-use equipment, that as yet are barely foreseeable. PV would ultimately become embedded not just as a component in electricity systems, but also in the social and institutional structures associated with the pursuit of market efficiency and environmental sustainability.

Such transformations would take many decades to evolve. To accelerate the process, as much attention will need to be paid to the potential markets, users and related institutional issues as to the technology itself. Some utilities in southern Europe are already involved and are playing an important role in development of grid-support and (where they carry the responsibility) remote applications. But to capture the larger long-run potential, architects, building owners, urban designers and perhaps even relevant transport planning authorities would need to become involved and familiar with the possibilities – as would the architects of European electricity liberalization. Even with such assistance, the extent of changes required is likely to make the expansion of PV not only later, but also slower, than that of wind energy – at least in Europe. With direct government backing, PV could achieve the 'PV-2010' target of 2000 MW, which would generate over 0.1% of EU-15 electricity supply. The analysis of potential PV markets, projections and expansion issues in this chapter suggests that more market-led expansion could average about 15–25% per year over the subsequent two decades, based primarily upon building-sited applications. Combined with

---

[25] G. Helisher et al., 'Guaranteed solar results of photo-voltaic systems', in *Proc. ALTENER Conference on Renewable Energy Entering the 21st Century, Sitges, Spain, 1996*, leaflet and guidelines EC-DG XVIII (Brussels: 1996).

smaller contributions from isolated and grid support applications, this could lead to a contribution in the range 2–6% of EU-15 electricity supply by 2030. Given the resources, expansion would probably continue thereafter based upon additional applications, some of which probably cannot yet be envisaged. The complexity and uncertainty of PV technology and applications make it impossible to predict the likely scale of PV's ultimate contribution. But even without ever going to centralized power plant applications, the resource exists for PV to contribute substantially to European electricity demand for buildings, and more speculatively to Europe's transport needs, and on a much greater scale to the needs of developing countries.

5.10 Conclusion

The relationship between cost and value is the ultimate driver of market development. Photovoltaic costs and benefits are in rapid transition, and both are strongly dependent upon climate and siting details, for both grid and off-grid applications. Costs depend upon the quality of the local solar resources, but benefits may be even more site-dependent. In some parts of southern Europe, the value of photovoltaics can already exceed its costs in isolated applications and on islands. Likewise, where land is expensive or where power distributed to buildings has a high value, rooftop systems will grow in importance more rapidly. The market will grow with familiarity and technical development, through different market segments, as discussed throughout this chapter. But the size and pace of development depends on specific actions that are undertaken.

If PV is ever to make significant contributions to European energy supply, the analysis of this chapter suggests that the fostering of innovation still needs to be the guiding principle of policies: innovation in technologies, applications, systems and institutions. European resources are not among the best and the industry is not well placed to compete on the basis of sheer financial backing; so PV must compete on the basis of brain power. The *need* to develop smarter technologies and applications for Europe could be a driving force for global competitiveness. In that case policies must not undermine that pressure by simply subsidizing easier, more conventional applications; rather, policies need to build up appropriate innovations and capabilities. This chapter has surveyed the many and wide-ranging policy issues that PV raises in Europe, and these can usefully be summarized in terms of the five generic policy challenges set out in Volume I.

(1) *RTD policy*

Basic technological developments are still required and many avenues remain open for exploration, far more so than for wind energy. The JOULE-THERMIE programme has established itself as Europe's primary vehicle for supporting such developments and PV already attracts almost half the renewable energy budget. Improvements could be sought particularly in manufacturing engineering and the more advanced, thin-film possibilities, and in testing and demonstrating applications, so as slowly to foster experience and familiarity with advanced technologies and applications.

(2) *Internal market and industry issues*

The ALTENER programme is tackling issues associated with establishing common technical standards and other aspects related to enhancing trade and consumer confidence. A key question is whether and how EU efforts can also foster greater cooperation among the various companies, encouraging them to specialize in different aspects, and how any support for a scaled-up industrial production facility can be allocated (which could also be addressed through the IEA implementing agreements). The biggest dilemma is whether the relatively small financial resources available should go towards supporting scale-up of existing crystalline cell production technologies, or should focus more upon building up capabilities in ultimately more promising technologies. The argument for the former is that such support would both enable wider applications and increase the cash flows to enable existing companies to remain in the business while they explore other lines of development. The risk is that much of the public expenditure would be drawn towards a technology that is inherently limited to small markets, and it could entrench industrial interests in this rather than in thin-film technologies. Extraordinarily, neither the Madrid Action Plan nor the *Photovoltaics in 2010* study addresses this question; the assumption appears to be that expansion of PV industries is good in its own right, even if they remain based on crystalline technology.

The liberalization of the electricity industries itself has many implications for PV. Some utilities have been important in fostering PV developments and sponsoring markets; this will become much more difficult as the companies become more cost-conscious and are driven by short-term financial interests. The underlying need for electricity companies to join together in sponsoring long-term research is nowhere more apparent than in the context of PV, and European governmental institutions could be important in brokering this. On the more positive side, liberalization can aid PV if and as its structure deepens to reflect the benefits of embedded generation, and in particular if and as it is pursued to the level of smaller end-uses, as will occur in the UK in 1998. The outcome of that great experiment – particularly relating to commercial buildings – should be of extreme interest to the PV industry.

(3) *Market supports*

In fact, all the experience summarized in Volume I, Chapter 7, suggests that market supports are a more effective way of stimulating innovative industrial developments than subsidizing production facilities. In the case of PV they would cause industries to focus on the whole chain from production to users, making business and marketing plans accordingly and fostering the institutional relationships that would be required for sustaining sales. The problem for PV is that such supports appear harder to target than for wind energy; they may need to be differentiated according to the different applications. Electricity distribution companies could receive capital grants for grid-supporting PV applications; transparency in the analysis of the value of installations to the system would help to increase awareness and capabilities. However, with a few notable exceptions, utilities have not fostered rapid development of any renewable source; independent investments have generally led to faster growth, and this also accords with the direction of electricity liberalization. Yet independent generation for grid supplies – as in the case of wind energy – is, as noted, hardly a serious option; the opportunities arise in applications to end-users.

One approach could be to subsidize appropriate monitoring and two-way metering technologies combined with specific buy-back tariffs; these can record the value of electricity produced for on-site use together with any sold back to the distribution system. PV generation used on-site would presumably be exempt from the sales taxes that now apply to electricity in most of Europe, and the approach would foster the technologies and concepts required for implementing distributed generation down to the level of end-uses. End-users could similarly receive capital grants; but these alone could distort the structure of investments, because they do not particularly encourage applications designed to maximize performance and output; grants per kilowatt of installed capacity would encourage high nominal capacity more than appropriate siting and reliable performance. Unfortunately, since end-use applications by definition are not targeted primarily at selling electricity back to the system, output credits *per se* may not encourage development of such applications. Inclusion of PV as a separate band in a system of renewable 'portfolio standard' requirements (Chapter 8), with end-use PV applications generating credits for the electricity supplier, could be the most effective approach.

(4) *Internalization of environmental costs*

As noted above, PV costs are too high for politically credible levels of internalizing costs associated with fossil-fuel emissions to be the main plank of a policy supporting PV (though it would help). PV in particular would benefit from moves that internalize the environmental impacts of transmission and distribution – for example, relating to power lines over mountains. Currently these are effectively cross-subsidized by most utilities. Achieving greater cost transparency in distribution, and evaluation of the external impacts of these activities, would be important elements in developing distributed generation which weights the local environment more heavily, thereby encouraging on-site generation. The existence of such cross-subsidization is a clear rationale for stronger support for installations at remote sites and for island generation.

(5) *Integration with other policies*

Perhaps the biggest challenge of all for PV is its integration with other policy areas. Because the most economic applications are directly tied to end-uses, effective policy needs to be integrated with those sectors. If PV is really to develop as an application for buildings, its use needs also to be built into courses for architects, builders and electrical systems engineers, and into the valuation of properties by estate agents. If it is to be developed for DC applications, that means a whole new course of training for electrical engineers and the involvement of appliance manufacturers. For the putative transport applications, PV options would need to be placed in the context of the ongoing vociferous debate on transport policy.

The greatest need of all, given the nature of the global resource, is to integrate PV options with national and EU development and export efforts. As noted above in the discussion of the global context, this is far from easy; yet it must be an important element in exploitation of the world-wide potential for PV and acceleration of cost reductions through the growth of total sales.

Chapter 6

Ocean Energy from the Tides and Waves

Tidal energy in estuaries, tapped by large dams, has attracted study for nearly a century but little has been exploited. The biggest resources lie in a few huge potential schemes in the UK and France that could each represent about 5% of national electricity production. Such schemes would attract strong environmental opposition, though in some respects the impact might be positive. The very high capital cost and long lead times of these giant schemes make their economics very sensitive to financing conditions, which are far more difficult in a liberalized system. These two factors, combined with the dominance of nuclear power in France, mean that such grandiose schemes will not proceed for the foreseeable future.

Much smaller tidal dams are possible. These are essentially a matter of local public choice because of the many and varied local impacts, both positive and negative. Smaller schemes represent an important local issue but a negligible contribution to European electricity production.

The principal alternative to tidal dams is to extract energy directly from tidal streams, especially where these flow through relatively narrow channels. Strong currents are known to exist around the UK and Greece, and to a lesser extent around France and Italy, and in the southern North Sea. The technology is insufficiently explored to allow credible estimates of costs and exploitable resources. The reduced environmental impact (as compared with dams) and the potential for more incremental development of technology and capacity justify further exploration of this option.

Wave energy is a major resource along the Atlantic continental shelf, with a technical potential in the range 5–10% of European Union electricity consumption; the fact that 70% of the energy comes in winter months would increase its value relative to conventional power sources. However, wave energy development has suffered from an acrimonious history in the UK, and the lack of a sustained research community means that cost estimates remain bedevilled by uncertainties and contrasting views. Estimates of 5–10 €/kWh have been cited for production-scale applications of some recent prototypes, with savings arising from combination with wind energy. There are opportunities for cost reductions, and wave energy cannot be rejected as unviable on the basis of current knowledge.

However, it is unlikely that offshore renewables for bulk supply can compete unaided against gas power generation without big gas price rises; their development must be considered in the context of diversifying electricity sources for long-term stability.

The approach taken to wave energy assessments in the UK has embodied two intellectual flaws: the failure to consider carefully the economic and energy circumstances in which wave energy might actually be deployed, and the failure to recognize that the relevant policy question is how best to develop information, expertise and capabilities.

Availability of undersea transmission cables would improve the economics of wave energy. The North Sea resource needs to be reconsidered in this context, as do the possible implications of a connection to the Icelandic renewable energy resources. Policy thus needs to be driven by RTD and small-scale trials, but in the context of considering complete systems.

The policy issues are similar for all the offshore renewable energy technologies, for which a two-track strategy appears appropriate. The first requires detailed and sustained study of recent and potential advances in marine technologies, of potential synergies between different offshore technologies and infrastructure established for different or multiple uses, and of the possible value of the electricity generated in different economic and energy circumstances. The second track involves incremental development of an offshore renewable energy community based around iterative development and assessment of pilot projects, using policies analogous to those used at the outset of the Danish wind energy programme. Shoreline and near-shore technologies, though very limited in terms of total electricity contributions, could play an important part in this learning process, especially in favoured locations (e.g. for islands or remote supply).

6.1 Introduction

The oceans contain and concentrate large quantities of renewable energy. The nature of the resource and the marine environment mean that these resources can be difficult to exploit, and little development – even of experimental schemes – has occurred to date. Yet marine technology itself has developed hugely, and the possible need for a crowded continent like Europe to exploit its renewable resources to the full during the next century raises the question of whether and how to approach the development of oceanic renewables. This chapter explores briefly the potential and policy issues arising.

6.2 Tidal power dams

Extracting energy from the tides can be done in two principal ways. The main candidate involves placing what amounts to a large, low-head, hydro dam across an estuary. The other is the tidal equivalent of 'run-of-river' hydro schemes, for tapping tidal streams.[1] The former involves mostly conventional and familiar technology, though in an application which is novel with the exception of the 30-year-old La Rance tidal barrage in France and some small-scale and experimental schemes.[2] Tidal streams are more speculative at present.

The tidal rise in the open seas is about 0.5 m; at the coast, this is amplified by the effect of shallow water, funnelling, and resonance effects, according to the characteristics of the estuaries. The energy available increases as the square of the tidal height and it is widely accepted that tidal power is only of potential economic interest where the tidal range exceeds 5–6 m. The physical resource in Europe is shown in Figure 6.1.

---

[1] A third approach to tidal energy has been proposed, which is to tap the energy from the rise and fall of water in the open seas, using floats. The *New Scientist* gave good coverage to a Spanish proposal involving piston tanks 200 m below the surface, connected to floats located 'about 8 km from the coast, and covering 5000 m^2 for a 1 MW power station' (*New Scientist*, 2 April 1994). The present author finds the approach and its claimed economics implausible, but like all such proposals it could usefully be subject to a simplified, open process of technology assessment.

[2] La Rance generates about 0.5 TWh/yr from an installed capacity of 240 MW. Like most tidal schemes and proposals, it was designed to generate electricity from the ebb (retreating) tide, but can also generate on the flood tide when appropriate.

Figure 6.1 Tidal heights in northwestern Europe (metres)

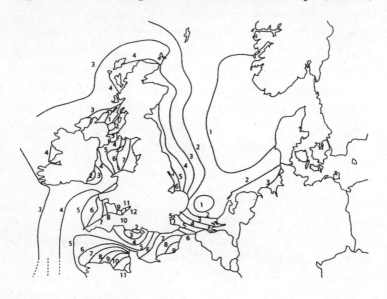

Source: J. Cavanagh, J. Clarke, R. Price, 'Ocean energy systems', in T.B. Johansson et al., *Renewable Energy* (Washington DC: Island Press, 1993).

The sites of possible interest are almost all confined to the west coast of the UK and northern and western France, though there are some possibilities in Ireland. Table 6.1 shows costs and output data for sites in Europe that have been estimated to offer output exceeding 0.1 TWh/yr at a cost below 7.5 ∈/kWh at a 5% discount rate. The resource on these criteria, around 40 TWh/yr, amounts to about 2% of EU electricity demand, but this is dominated by two huge potential schemes: the Baie St Michel in France, and the Severn Estuary in the UK, each of which would supply about 5% of those countries' electricity requirements. In the UK the Solway Firth offers another potentially very big scheme, and Strangford Lough offers significant power for Northern Ireland at somewhat higher costs.[3]

---

[3] Feasibility studies by ETSU suggest output and costs from these schemes to be 10 TWh/yr and 0.53 TWh/yr, at 8.7 and 12.5 ∈/kWh, respectively. For both schemes, the more general parametric modelling approach used for many of those listed in Table 6.1 shows higher output and lower cost, which would have brought them within the scope of the table.

Table 6.1 The most significant potential tidal estuary schemes in Europe[a]

Country and site	Installed capacity (GW)	Annual energy (TWh)	Generation cost (in 1989 and [1995][b] ∈/kWh)			
			Low discount rate[c]		High discount rate[c]	
France						
Golfe du Morbihan	0.37	0.66	4.8	[5.9]	8.5	[10.5]
Seine	0.29	0.56	5.2	[6.4]	9.2	[11.3]
Baie Mont St Michel	7.91	18.67	5.5	[6.8]	10.8	[13.3]
Baie de Somme	0.77	1.64	6.0	[7.4]	11.9	[14.6]
Pertuis Breton	0.96	1.76	6.4	[7.9]	11.8	[14.5]
Etel	0.064	0.114	7.0	[8.6]	12.5	[15.4]
United Kingdom						
Severn	8.640	17.0	5.1	[6.3]	10.8	[13.3]
Fleetwood/Wyre[d]	0.056	0.131	5.2	[6.4]	9.9	[12.2]
Hodbarrow/Dudden	0.16	0.31	5.9	[7.3]	10.5	[12.9]
Liverpool/Mersey[d]	0.70	1.5	6.0	[7.4]	12.2	[15.0]
Taw/Torridge	0.10	0.19	6.4	[7.9]	11.5	[14.2]
Other						
Castlemaine (Ireland)	0.102	0.181	7.0	[8.6]	12.5	[15.4]

[a] The table shows potential tidal schemes with estimated output exceeding 0.1 TWh/yr at a cost below 7.5 (1989) ∈/kWh at 5% discount rate. Other big sites, at only slightly higher costs, include the Solway Firth (10 TWh/yr) and Morecambe Bay (5.3 TWh/yr) in the UK, and Strangford Lough (0.5 TWh/yr) in Northern Ireland.

[b] 1989 ecus have been recalculated to give 1995 ecucents using the consumer price index.

[c] Low discount rate is 5%, high discount rate is 10%.

[d] Data derived from detailed site feasibility studies. Others are derived from rough parametric modelling.

Source: ETSU/CCE for the European Commission, DG-XVII (Brussels: undated).

Tidal power offers sustainable energy and the operating costs are negligible. Yet the biggest concerns about tidal power are its local environmental impacts and costs. The environmental objections concern complex possible impacts on bird life, fish, and sediment flows and deposition, but can be simply encapsulated: tidal schemes would interfere with unique estuarine ecosystems. The environmental response is equally simple: there is remarkably little hard evidence that tidal schemes would actually be damaging. Indeed, most detailed studies suggest that they would benefit the richness of the estuary ecosystems by reducing peak flow rates and upstream salinity, and increasing the duration of high tide. The debate between precautionary instinct and analysis based upon inevitably incomplete understanding is not one that is likely to be readily resolved.

An important aspect of tidal developments is that they generally increase the amenity value of the area. The big tides and rapid flows in estuaries that are of interest to tidal power make them poor places for most forms of water recreation; tidal barrages would make them more amenable and could themselves be a notable tourist attraction. They also provide an estuary crossing, the value of which depends upon the location. In some cases such benefits may even outweigh the energy generation aspect. A tidal barrage, in other words, is not something that can be considered independently of its broader socio-economic implications for the region.Nor indeed should tidal power be considered independently from other renewables. Tidal barrages could offer natural foundations for wind turbines capturing strong coastal winds, as they have already been installed on some Dutch levees. More speculatively, integrating wave energy converters in the face of dams during construction may be sufficiently cheap to justify this (probably small) energy addition. Both would help to produce smoother output than tidal energy alone and utilize the mechanical and electrical infrastructure more efficiently.

The final important aspect is the capital structure. Like hydro plants, the cost is almost all up-front, and the energy may then flow at very low operating cost over a century. Evaluated at private-sector rates of return, and excepting a few small schemes justified largely on non-energy grounds, even the best sites look hopelessly uneconomic. Evaluated at very low social rates of return, they offer a remarkable opportunity for low-cost, clean energy throughout the next century.

However, the two largest schemes, the Severn crossing and the Baie Mont St Michel, would each involve construction over periods of up to ten years at costs running into tens of billions of ecus and would also attract massive environmental opposition. Despite repeated and exhaustive studies of the Severn scheme, it seems little closer to implementation than when first proposed almost a century ago. With the UK's surplus of generating capacity and financial stringency, the UK government seems most unlikely to become involved in backing the Severn scheme for the foreseeable future

– and the experience of the Channel Tunnel is hardly likely to endear private investors to the idea. France, with its surplus of nuclear capacity, seems even less likely to back the Baie Mont St Michel project. The liberalization of electricity supply, effectively removing the possibility of firm, risk-free markets for the power generated, is probably the last nail in the coffin of such grandiose schemes. A banking consortium financing tidal power in a liberalized market could end up extremely rich – but probably only after all the people involved were long dead.

The situation for small schemes is less clear-cut, because the scale of finance and lead times is not so large, and because local authorities could raise the finance for local returns. Nevertheless the combination of environmental, local-amenity and financial characteristics indicates that tidal power is quintessentially a matter of public choice. A decision whether to build has to be taken by local and national government, they have to finance it and they, the local community and the global environment would receive the benefits – with the local community also having to offset this against the local environmental impacts. So far, public choice has not come out in favour of tidal energy schemes, often because more wide-ranging objections by local conservationists have won out against the national benefits and support of local authorities.[4] Because this is a matter of local and national preference and economic philosophy, and because the major resources are located in just two EU countries, there is little role for the EU except possibly under the umbrella of broader financial strategies discussed in Volume IV of this study.

6.3 Tidal stream energy

An alternative to building dams across estuaries is to extract energy directly from tidal flows, in much the same way that wind turbines obtain energy from the wind. Fraenkel has summarized the advantages of this approach, compared with tidal dams, as including the more widespread nature of the resource, the steadier nature of the output (particularly by combining different sites), the modular nature of the technology and the lack of any coastal environmental impact.[5]

---

[4] For example, 'Welsh barrage plan is rejected', *Financial Times* (14 September 1995).
[5] P.L. Fraenkel, *The exploitation of tidal and marine currents* (Eversley, Hants, UK: IT Power Limited, 1995).

Table 6.2 Principal identifiable tidal stream sites

Site	Mean velocity (m/s)	Mean depth (m)	Energy available (GWh/yr)
Scotland:			
Dorus Mor & Kyle Rea	1.5	15	790
Pentland Firth (6 sites)	1.6–2.2	57–76	10,000
Faroe Islands	1.4	25	85
Orkney – Westray Firth	1.3	25	135
Shetland (3 sites)	1.3	25–35	285
English Channel:			
Portland Bill	1.5	29	652
Channel Islands – Race of Alderney	1.7	34	6,470
Channel Islands – Cap de la Hague	1.3	53	2,351
Mediterranean:			
Aegean – Gulf of Caloni, Lesvos	2.0	n.a.	n.a.
Aegean – Samos Strait (2 sites)	1.8	35–60	2,635
Strait of Messina, Italy (3 sites)	1.4	60–110	85
Total			**23,400**

One option is indeed to use axial-flow rotors, on the same principle as horizontal wind turbines. A consultant's report for ETSU examines the resource around Britain and the costs based upon this approach.[6]

In 1996 a study for the European Commission completed a more detailed appraisal of the technologies and European resource.[7] It estimated the cost of electricity from a basic axial-flow turbine (like an undersea wind turbine) to be 0.07 ecu/kWh in an exceptionally fast current of 3 m/s, and suggested simply that sites with current speeds 'in the range 1.0–1.5 m/s are much less attractive in terms of energy capture' (as with wind energy, the energy density increases as the cube of the current speed). The crudeness and preliminary nature of the design means that cost estimates are speculative at this stage, and there could be considerable scope for improvement. Also about half the costs were estimated to be associated with support structures and sub-sea transmission, which in some circumstances could be reduced by using existing infrastructure as indicated in Chapter 7.

---

[6] ETSU, *Tidal Stream Energy Review*, ETSU T/05/00155/REP (Harwell: ETSU/DTI, 1993).
[7] *The Exploitation of Tidal and Marine Currents*, EU Contract JOU2-CT93-O355 (February 1996).

Table 6.3 Wave power intensities and technical potential around Europe

Area	Country	Average deep-water wave-power level (kW/m)	Technical potential	
			(TWh/yr)	(% of current generation)
North Sea	Denmark	18–20	5–8	15–25
West coast of Ireland	Ireland	47–52	21–32	120–200
and UK	UK		43–64	13–21
West coast of France	France	29–32	12–18	3–5
and northern Iberia	Spain		10–16	7–12
West coast of southern Iberia	Portugal	24–26	12–18	40–70
Mediterranean Sea	Italy	11–13	9–16	4–7
Greece			4–7	12–20
Total	**EU-12**		**116–179**	**6–9**

Source: ETSU/OPET, 'An assessment of the state of the art, technical perspectives and potential market for wave energy' (ETSU/CEC, 1993). Derived from Tables 1 and 2.

The study reviewed over a hundred possible sites and Table 6.2 summarizes sites with average current speed of 1.3 m/s or higher. The best sites are around the UK and Greece, with others of possible interest close to France and Italy. The electricity potentially available from these sites amounts to about 1% of EU-15 electricity supply. Though the study only examined coastal sites, undersea topography can also form strong currents; currents peaking at around 1.5 m/s are known to exist in the southern North Sea, as well as offshore around some of the western Scottish islands.[8]

Clearly, extracting energy from tidal and other ocean currents is an option still at the earliest stages of appraising technology and resources. From a policy perspective, in addition to its probable very low environmental impact, the attraction of tidal stream energy is that an incremental approach to technology development is possible, exploring a wide variety of ideas for modest outlay, and moving on to start testing out on a small scale any that cannot obviously be rejected. Resource surveys combined with a more substantive technology appraisal would seem justified to initiate this process, but at present little more can be said about the realistic prospects for tidal stream energy.

[8] Shell UK, personal communication.

6.4 Wave energy: the European resource

Europe lies at the latitudes of maximum ocean wave energy intensity. The intensity of the resource in the seas around Europe has been assessed for the European Commission, with the results shown in Table 6.3. The Atlantic region west of Ireland and Scotland has the most intense resource; the intensity declines going south, and even more going into the North Sea.

The crude technical potentials set out in the table include technical constraints upon practical conversion efficiencies of devices, but take no account of these differing intensities or economic, competing sea-use or environmental constraints. For Ireland, the technical potential is comparable with total electricity demand, and it is also relatively large for Portugal. The UK has the biggest absolute figure, which represents 13–21% of the UK 1990 demand. The technical potential for Norway is also large, and some assessments give somewhat higher figures.[9] In all, this points to a technical resource probably in the range 5–10% of EU-15 electricity demand.

Closer to land, the resource is reduced by friction, at the seabed and by declining wave heights. The Commission report estimates the resource at potentially more easily exploitable depths of 15–25 m to be about one-third of the deep-water resource.[10] The resource at the shoreline is much smaller still, and is further constrained by a range of environmental factors, though the economics may be considerably improved; the UK estimates the shoreline resource at costs below 10 p/kWh (equal to about 13 €/kWh) to be about 0.4 TWh/yr, almost all in Scotland.[11]

Wave energy is a variable resource, but in a very useful manner for northern Europe: most of the energy is concentrated in winter months (about 70%

---

[9] Technical studies differ; D. Mollison ('The European wave power resource', in L. Duckers (ed.), *Proc. Conf. on Wave Energy Devices* (Coventry: Solar Energy Society, 1989) suggests that the maximum technically achievable resource could be as high as 400 TWh/yr using highly efficient devices, more than twice the maximum figure in Table 6.3. Salter has subsequently estimated the total theoretically exploitable resources from Iceland to Cape St Vincent to be about 600TWh/yr (evidence to meeting on small hydro, DG-XII, Brussels, September 1996; personal communication).

[10] ETSU/OPET, *An assessment of the state of art, technical perspectives and potential market for wave energy* (Harwell: ETSU/CEC, 1993).

[11] DTI, *The UK's Shoreline and Nearshore Wave Energy Resource*, ETSU WV1683.

Figure 6.2 Classification of wave energy devices

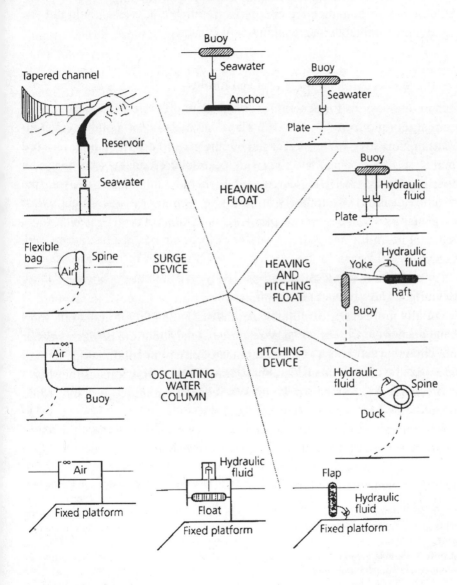

Source: ETSU R82.

of output would be in the six months October–March)[12] and during most of
that period the load factor could be quite high, with few periods of very low
output. Per unit output, wave energy from reliable devices should thus be
slightly more valuable than conventional capacity.

6.5 Wave energy technology and economics

Technology designs for generating power from offshore wave energy have
been largely pioneered in the UK. There have also been notable developments in Norway, Sweden and Japan but the majority of these have focused
upon shoreline devices, and their limited offshore concepts do not seem
more promising than UK designs although they add interestingly to the
range of ideas.[13] Unfortunately, in terms of technology assessment, wave
energy has proved the most complex and contentious of all renewable energy
technologies in Europe. A huge variety of technologies has been proposed,
sometimes with zeal, in the face of a long history of official assessments
which state these devices to be hopelessly uneconomic.[14] Some of these
technologies are shown in Figure 6.2.

Virulent controversy in the UK over the economic assessment of these
designs dates back to the early 1980s, when the decision to abandon significant UK government support for wave energy was justified with reference
to a report by consultants RPT indicating extremely high costs, in the order
of 20–50 p/kWh. One of the highest-cost devices was the 'nodding duck'
pioneered by Stephen Salter, and Salter led a campaign which finally led to
an admission in 1990 – following an inquiry by the UK government's Select
Committee on Energy – that the 1982 assessment contained both typo-
graphical and substantive analytical errors in the numbers presented, including implausibly low levels of availability for the transmission of electricity to the coast and some other components.

---

[12] Ibid.

[13] A brief review is given by Tony Lewis, in *Wave Energy: Current Research Activities and
Recommendations for European Research Programme*, Report to CEC DGXII (Brussels:
1992/3).

[14] For a lively (and angry) account of the ideas and of the sorry history of wave energy technology assessment, see David Ross, *Power from the Waves* (Oxford: OUP, 1995).

Table 6.4 Generation costs of main wave energy devices estimated by UK 1992 Review

Device	Electricity cost (p/kWh)[a] and [1995 ∈/kWh][b]			
	Low discount rate (8%)		High discount rate (15%)	
Bristol cylinder	12	[20]	20	[33]
Edinburgh duck	16	[26]	26	[43]
NEL OWC[c]	16	[26]	29	[48]
SEA clam	8	[13]	12	[20]
Shoreline OWC[c]	6	[10]	9	[15]

[a] Median costs at 1990 prices, assuming successful completion of outstanding RTD, sitting in North Atlantic.
[b] Recalculated using the consumer price index.
[c] OWC, oscillating water column.
Source: T.W. Thorpe, *A Review of Wave Energy: Vol. I, Main Report*, ETSU-R-72 (Harwell: ETSU/DTI, December 1992), p. viii.

Following this, the UK government established a fresh investigation into wave energy resources and economics, which represents the most thorough analysis to date.[15] The assessment was devised as a consensus-building operation, with participation from the project teams and guidance by an independent committee, and in this respect largely achieved its purpose, gaining commendation from all sides. The total UK technical resource was estimated to be 10 GW mean power (e.g. 40 GW at 25% load factor, yielding 87 TWh/yr) in deep waters, or 7 GW in shallow waters; the former is somewhat bigger than the Commission estimates in Table 6.3. The study investigated representative designs in five categories of wave energy device. The estimated costs of each device are summarized in Table 6.4.

These costs still indicate wave energy to be substantially more expensive than conventional power at current fuel prices, particularly for the offshore devices where the costs are three to six times those of conventional power generation, even on assumptions about the success of RTD needs identified by the assessment. But huge uncertainties remain. The review was assessing

---

[15] T.W. Thorpe, *A Review of Wave Energy: Vol. I, Main Report*; ETSU-R-72 (Harwell: ETSU/DTI, December 1992), p.viii.

specific designs proposed as of about 1990, none of which had undergone any sustained critical review and development. Two-thirds of the potential electricity was still lost in power collection. In the analogy of Stephen Salter, it was a bit like trying to assess the economics of flying at the time of the Wright brothers. Furthermore, the assessed costs would decline substantially for lower discount rates (not as rapidly as for hydro power because of the higher operating-cost component).

Thus the debate was far from over. For example, prompted by the insights generated by the wave energy review process, Salter undertook a fundamental redesign of the duck, with results which are best described in his own words:

The central cost estimate of the 1992 Thorpe review is 16 p/kWh – very close to the cost of wind energy in California in 1985 ... the effect of ... [basic design changes] ... should reduce the capital cost by a factor of about two ... the extra productivity of complex-conjugate control should give us an extra 1.5 times the energy at the generator terminals ... the development of methods to send electricity reliably along the spine during productive months should deliver about twice the energy to land. All we need now is a decision-maker who can accurately divide.[16]

The resulting cost estimate of 2.6 p/kWh[17] – or a more widely quoted figure of 4.0 p/kWh arising simply from the revised capital cost and transmission assumptions – could bring this within the range of serious policy interest. The problem for the decision-maker is not division, but the credibility of estimates. The official review focused upon concepts that had been developed largely in a vacuum. Obviously, a revision that so radically reduces energy costs is subject to scepticism, but was not evaluated in the published Thorpe review. The almost non-existent funding for the pursuit of wave-energy ideas over the preceding decade left no basis of considered review and revision upon which to try to build consensus.

---

[16] Stephen H. Salter, 'Changing the 1981 spine-based ducks', in *Proc. IEE Conference on Renewable Energy: Clean Power 2001*, Conference Publication No. 385 (London: IEE, 1993).

[17] 2.6 p/kWh in 1993 is equal to 3.5 ∈/kWh (1995 ecucents); 4.0 pence to 5.4 ecucents.

To the surprise of many, the next stage in the saga was taken by the private development of a 2 MW prototype of a different device, the ART OSPREY oscillating water column for near-shore generation. Another assessment of this specific device, in conjunction with the design team, suggested that the cost of electricity should be about 6.4 p/kWh (at 8% discount rate – 7.7 € /kWh) both for small-scale near-shore deployment and for a massive scheme for deeper waters.[18] However, it was also noted that the cost of energy could be reduced, to about 5.1 p/kWh (or 6.2 € /kWh), by siting a 1.5 MW wind turbine on top of the 2 MW OSPREY. The study noted that the OSPREY is 'at the early prototype stage and requires further RTD to establish both its practicability and confidence in the cost and performance values'.

In 1995 the prototype, backed by a private consortium, was launched. Unfortunately it was damaged at launch and sank in heavy seas while the ballast tanks were still being filled. Undeterred, the backers said that a second machine would be developed quickly, and it is due to be installed (and covered by insurance) in 1998 at costs estimated in the region of £1,000/kW (or 1,200 ecus/kW).[19] Despite the setback, the growing interest in OSPREY and similar devices does suggest a more incremental technical approach to wave energy, with near-shore generation forming an important intermediate step, and resource, on the path towards deep-ocean devices.

Despite the advances, at present the prospects for wave energy can hardly be described as propitious. In some respects the technical assessments are optimistic in their assumptions that open RTD questions will be resolved. Also, the fact that the bulk of wave energy lies at the west European periphery, far from areas of high electricity demand, means that its harnessing would entail significant onshore transmission costs. But there are mitigating factors. One is simply the financial basis of the calculations – the UK government's dogmatic insistence on placing the emphasis upon energy cost estimates at an 8% discount rate, for example, seems faintly ludicrous for reasons set out below. Moreover, not only are design developments –

[18] T.W. Thorpe, 'Assessment of the ART OSPREY wave energy device', ETSU-R-90, Final Report (Harwell: ETSU, 1995).

[19] For a detailed account see 'The ups and downs of wave power', *NATTA Newsletter*, No. 98 (Milton Keynes: Energy and Environment Research Unit, Open University, November/December 1995).

and fundamentally new designs – still being advanced, but there must also be some scope for reducing the costs of basic components. The development of strong, resilient, lightweight composite materials at lower costs for such bulk applications would be particularly important. Furthermore the TERES chapter on wave energy[20] claims that more advantage could be taken of expertise and facilities available from the conventional marine industries, and criticizes the wave design teams for failing to do so; however, industrial participation has been growing in some of the design teams and development efforts, often after considerable initial reluctance.[21]

Other potential complicating factors are the dependence of costs upon wave intensity and the importance of transmission costs and reliability in the economic assessments. Many of the costs are determined by the exacting requirements of standing up to, say, once-in-a-century peak waves, and the output also has to be limited in strong seas. Economic assessments of offshore systems have focused upon the deep Atlantic, but it cannot be assumed that generation in the North Sea would be twice as expensive just because the average wave intensity is halved. The relationship could be very much weaker, and this is important because sites in the North Sea could both reduce transmission costs considerably, and generate energy much closer to the southeast of England where it is needed, and/or to the Danish and Continental power networks where electricity costs may be higher. An additional possibility, discussed further in the next chapter, is to link into sub-sea transmission established for other reasons; transmission costs appear to form 10–30% of the estimates in the *1992 Review*. The uncertainties still appear large enough to mean that wave energy cannot simply be rejected as inherently unviable.

6.6 Wave and tidal energy policy issues and strategies

The policy issues for *shoreline* wave energy and estuary tidal energy (excluding the big schemes discussed above) are relatively straightforward. The resource, as constrained by environmental and other factors, is so small

---

[20] *The European Renewable Energy Study*, Vol. II, EC-DG XII (Brussels/Luxembourg: 1991).
[21] Stephen Salter, personal communication.

and the costs sufficiently high that in themselves they are not of strategic interest for grid-based power production. Conversely, the output is big enough and the costs sufficiently low for them to be of considerable interest for supply to isolated islands, where the primary concern is adequate and affordable energy. The JOULE–THERMIE programme could be (and has been) involved, especially to the extent that some schemes may give useful information for offshore devices; apart from that, installations could also draw on structural funds, particularly VALOREN.

Moving offshore, the resources are much bigger but so are the difficulties – technically, and for policy. The unique feature of ocean-based renewable energies is the timescale involved and the scale of the uncertainties in almost all directions. Tidal barrages would last through the twenty-first century, and tidal current schemes and wave energy devices would need to undergo perhaps decades of development and sea trials for a large-scale industry to become established and move towards full-scale deployment. Not only are the technical uncertainties large, but so is the economic and environmental context. It is, for example, a highly debatable assumption that by the middle of the next century investment opportunities will be available at an 8% real rate of return, or that the costs and criteria for assessing electricity supply will remain focused on discounted cash flow analysis compared against current costs of a few ecucents per kilowatt-hour.

Furthermore, offshore energy comprises three independent options: energy from waves, tidal currents and wind. In each case, two dominant cost factors are transmission of the power to shore, and moorings or foundations. Development of two or even all three options together might be able to share these basic infrastructural costs, maximize energy output, and produce a higher load factor than would be possible from any individual source. Assessment of the OSPREY device highlighted cost advantages of combining wind and wave power, and Fox has argued that there are more fundamental engineering advantages arising from the distribution of mechanical loads and reactions in combined systems.[22] In addition, the possibility of an

---

[22] See 'The ups and downs of wave power', op. cit. (Note 19), and various other issues of the *NATTA Newsletter*. As with much else in the field, the claims remain frustratingly unexplored in the absence of a more established and critical community.

expanding network of sub-sea cables, existing infrastructure associated with the oil and gas industries, and the economic transformation of the existing offshore industries – discussed further in the next chapter – all seem relevant to a serious assessment of strategies towards offshore energies.

The agenda of analysis has for too long focused upon national studies of particular devices proposed by clever inventors. The cost of energy has been too high, and research funding curtailed. This is a mistaken approach. The offshore oil industries have achieved cost reductions themselves in the past two decades that few thought possible, from an already established technical base: had the low oil price been forecast correctly and assessment been made in the way it has for wave energy, these developments might never have started. The approach taken to ocean renewables has similarly vastly underestimated the uncertainties, and the scope for change and for learning.

Specifically, the approach taken by the UK government towards assessing offshore renewable energy to date has rested upon two fundamental intellectual flaws. The first is trying to assess technology proposals against current conditions without detailed consideration of the possible circumstances in which offshore renewable energy might actually be deployed. The second is focusing assessment on an all-or-nothing choice of whether to pursue a specific technology, when in fact the only decision that needs to be taken at present is how to develop most efficiently the most relevant information, expertise and technical capabilities.

Consideration of these two dimensions of the policy problem points towards two branches of a strategy for pursuing offshore renewable energies that could be pursued by the UK and other interested parties.[23] The first is to

---

[23] The distribution of the resources introduces institutional complications. Ocean resources are certainly not spread well throughout Europe; but nor are they restricted to the UK alone. The EU's THERMIE programme rejected any involvement with offshore wave energy on the grounds that it is not nearly ready for commercial demonstration. A report to the RTD Directorate recommended an initial research programme of 1.2 million ecus to identify priorities (Lewis, op. cit., Note 13) but the RTD Directorate is understandably reluctant to spread its resources on such a contentious, geographically narrow and relatively less promising field compared with the major onshore resources. Development should probably be led by the UK, as the country with the largest resource and most wave energy pioneers, but could involve broader international collaboration among interested parties, under the aegis perhaps of the EU but more likely of the International Energy Agency, given the significant Japanese interest in the field.

understand better the possible contexts of deployment. What is needed are collaborative 'definitional' studies that explore in more depth the set of generic issues raised above, such as:

● the nature of established experience and infrastructure in the marine construction industries, their prospects and their possible relationship to the offshore renewable energies;
● the scope for lower-cost advanced materials with characteristics relevant to wave energy (weight, strength, flexibility and corrosion resistance) and their likely costs under conditions of mass production;
● the possible relationship between the different offshore renewables, specifically the scope for synergies between wave, wind and possibly tidal current energy;
● the potential scope, relevance and impact of multi-purpose sub-sea cables;
● appropriate economic criteria for assessment, including discount rates, that might be appropriate for assessment decades hence;
● the dependence of wave energy costs on different resource intensities, given re-optimization of designs;
● the value of wave energy to different power systems over time, given the seasonal characteristics;
● criteria for taking RTD forward to device development, including approaches to assessing the likely costs of development and demonstration programmes for ocean-going devices.

Such studies, by their nature, should have a broader compass than the wave energy community alone; they should incorporate electrical-system and marine engineers, and specialists in the economics of energy systems and environment.

In the second dimension, effort needs to be directed not solely at assessing specific devices for prototype development but at the process by which knowledge and expertise can be accumulated. Although the 1992 Wave Review was ostensibly focused upon assessing the costs of specific devices, it showed signs of groping towards a better approach, for more important than its results was its role in developing a measure of trust and common understanding between proponents and assessors, and the generation of challenges that led design teams to reconsider and improve their proposals. Had it been followed up, it could have helped to form the nucleus of a healthier wave energy community. For what is really needed is the ocean energy equivalent of the origins of the Danish wind energy industry: modest, long-term governmental support for a community of researchers and private efforts, interacting with government research laboratories and challenged by their reviews of proposals; testing of designs by government facilities; and involvement of relevant existing industries. Support could be given, in parallel, to design teams for any further design development and small-scale experiments including wave-tank or lake/shoreline testing of concepts or components.

With these parallel streams developed over a period of a few years, it should then be possible to issue a call for tenders for prototype devices, based on the criteria established by the definitional studies pursued in the first branch of the studies. We might thus at last embark on an iterative process to foster a steady expansion of the reliability, scale and economic feasibility of devices for tapping Europe's extensive oceanic renewable energy resources.

Chapter 7

Integrated Renewable Electricity Systems for Europe

In most of Europe, wind, wave and small hydro power could all profit from their positive seasonal correlation with electricity demand; storage and load management would also increase their value. In some cases, particularly in southern Europe and some service-sector building applications, PV also has a positive correlation with specific end-uses.

Information and control technologies combined with power electronics have improved the scope for integrated applications of renewables for localized supply. In addition to isolated supplies and specific end-use PV applications, four principal classes of grid-connected niche markets can be identified: islands, coastal regions, mountain village regions, and rural distribution, of which the first three are also often characterized by good primary renewable resources. Application of integrated renewable systems in these conditions is, however, inhibited by at least four institutional obstacles:

- *institutional weaknesses including the difficulty for the relevant communities to know their options and keep pace with technological change;*
- *the industrial fragmentation of the different renewable energy industries;*
- *the institutional structure of the existing electricity supply industries; and*
- *the widespread use of fixed national tariffs for domestic electricity supplies.*

The first of these obstacles may be overcome by including renewable energy in regional development programmes. The second has to be addressed primarily by the renewable energy industries themselves. Liberalization of electricity systems can start to address the last two, though allowing regionally differentiated tariffs will be politically highly contentious.

Advances in long-distance DC transmission technologies also improve the prospects for generation from remote concentrations of renewable energy, transmitted to European centres of demand. Liberalization, by enabling different industries to finance and develop the cheapest resources to supply some of the higher-priced electricity markets in Europe, could aid such developments, though the higher cost of capital could offset this.

The development of thermal solar power (probably solar thermal), hydro, and wind power in north Africa could ultimately lead to exports to southern Europe via Corsica, Crete or Spain. The extensive use of European finance for energy developments in these regions could be reviewed in the light of these possibilities.

For northern Europe, integrated systems for the North Sea and Atlantic could ultimately emerge from the development of ocean-based renewables and the transmission of surplus hydro and geothermal energy from Norway and Iceland. Using ocean-based renewables to supply power in the offshore petroleum industries, which embody tremendous knowledge of offshore engineering, could provide an initial entry market. Transmission and the costs of foundations are an important fraction of the costs of projected ocean-based renewables, but Norway is already laying power cables through the North Sea. At the same time the North Sea petroleum industries will increasingly have to address issues of decommissioning production facilities, which could in principle be used for siting or mooring of wind, wave or tidal stream energy devices if transmission exists to take the power. A plausible scenario could be as follows:

2000–2010. Development of offshore wind energy in the southern North Sea for supplying offshore electricity demands, perhaps also utilizing networked transmission and exploring prototype wave and tidal stream devices.

2010–2020. Development of central North Sea wind and wave energy, potentially utilizing transmission capacity associated with the export of surplus wind and other energy from Scotland and Norway and perhaps the Shetland Isles, possibly already including pilot tidal stream energy and/or transmission from Iceland to the Shetlands.

2020–2030. Completion of Icelandic exports, possibly including lines routed west to Ireland, associated with the development of the north Atlantic wave energy resource.

Such developments are speculative but the prospects are sufficient to justify a far greater collaborative effort to pursue relevant RTD and demonstration schemes that could draw together the wind energy industry, wave energy researchers and the existing offshore petroleum and engineering industries.

7.1 Introduction

Having considered each of the primary renewable electricity supply technologies, this chapter attempts to integrate insights from these specific studies with the big questions surrounding the structure and evolution of electricity systems outlined in Chapters 1 and 2: the interaction that will, to a large extent, determine the markets and returns available to renewable electricity, and hence their prospects.

This report has emphasized the very different characteristics of most renewable electricity technologies as compared with conventional supply options, and hence the importance of niche markets (which reflect local comparative advantages) and of other niche locations (which may reflect specific concentrations of low-cost resources, and/or use of existing infrastructure). Both provide a basis for larger industrial development, and arise in part from the dispersion of renewables and their relatively smaller scale. Also, given the variability of primary renewable electricity sources, their interaction with a system may be important. Consequently, this chapter considers the way in which different renewable electricity sources might combine in electricity systems on different scales – from the very small to the very large – and the policy issues that will determine the extent to which the opportunities are exploited.

7.2 Combinatorial characteristics: technical aspects

Before considering issues relating specifically to scale, some general characteristics can be noted about the way in which primary renewable electricity sources may combine, with electricity demand and with one another. Because they are all sources with variable output, their value is influenced by the correlation between the source and electricity demand and (at higher penetrations) by the extent of storage and load management options. Chapter 2 described the general principles relating to integrating intermittent sources, and the later individual chapters have noted how the characteristics of the source may affect its value in relation to the demand. In addition, different intermittent sources will interact in different ways, depending upon their own characteristics.

Table 7.1 Influence of correlation characteristics on value of primary renewable electricity sources

	storage	Demand	Hydro	Hydro	Wind	Wave	PV	PV	
			yes	noᵃ	no	no	yesᵇ	no	
Demand			++	+	+	+	−	—	
Hydro	yes	++		o	++	+	+	++	Seasonal
Hydro	noᵃ	+	+		−	−	+	+	correla-
Wind	no	o	+	−		−	+	+	tion
Wave	no	o	++	o	−		+	+	
PV	yesᵇ	++	o	+	++	+		o	
PV	no	+	++	+	+	+	o		

Short-term correlation

Note: The table illustrates the way in which correlation between the electricity demand (first row and first column) affects the *value* of different intermittent sources as compared with a baseload plant of equivalent output, and the way in which inter-source correlations affect the value of combinations. The top right-hand elements relate to seasonal correlations; the bottom left-hand elements relate to short-term correlations.
++ strong positive influence, + mild positive influence, − negative influence,
— strong negative influence
ᵃ Run-of-river hydro
ᵇ Short-term storage (batteries/flywheel).
Source: Author.

    Table 7.1 attempts to summarize how this may influence the value of the electricity produced, as compared with a conventional baseload plant in which the same average annual output is spread evenly throughout the year, and assuming moderate penetration of the system by any one renewable source (e.g. with each source supplying no more than 10–20% of the demand on the system, at the system scale considered). The relative value is influenced both by seasonal correlations (top right-hand half) and by short-term correlations, particularly during the season of greatest output (lower left-hand half). Tidal power variations have little short- or long-term relationship with demand or other renewable sources, and so (given that it is also a rare resource, irrelevant to most systems) tidal energy is omitted from the table.

    Thus, for example, the positive seasonal correlation between electricity demand and wind energy output in most of Europe increases its value, as compared with a source that has equal output throughout the year. The short-term correlation between wind and electricity demand appears slight, so the influence of this is small; but storage associated with hydro would increase the value of wind energy, and vice versa. Quantification is impossible except through extremely complex simulation of particular systems, but the sign of the influence is in most cases fairly clear from the underlying variations.

Wind, hydro and wave power are all seasonally correlated with electricity demand across most of Europe. PV output is generally negatively correlated against gross system demand, and this is one of the reasons why it is implausible that it could compete against centralized power generation. But ironically, in some of its 'niche' markets – particularly for on site supply to service-sector buildings – there may be a positive daily and/or seasonal correlation. In southern Europe there is potential correlation associated with air-conditioning and refrigeration loads. Even in northern Europe, refrigeration requirements are partly seasonal, and building electricity demand net of output from co-generation plants may well be positively correlated with PV. The value of such PV may also be significantly enhanced with a little on-site storage.

Thus, in much of Europe, most renewable sources gain economically from a positive seasonal correlation with electricity demand. In many cases, mixing different renewable sources, particularly if one of them carries some storage capacity, also has positive synergies. The 'bad weather' trio of wind, wave and run-of-river hydro are an exception to the positive synergy, since they are all inclined to yield maximum output over similar periods. Even to this there can be exceptions (e.g. if wind turbines shut down in storm conditions, the simultaneously greater output of hydro or wave power is an advantage): the presence of a combination of such renewables would tend to increase the value of storage and load management (and vice versa).

The importance of such factors should not be overstated. Hydro storage capacity in a system will not in most cases rescue the economics of wave energy using current technology, for example. Nevertheless, these factors should be relevant to marginal cases, or where the combination of different renewables can avoid the need for new construction of a conventional power plant or a long new distribution line; and they highlight the desirability of thinking in terms of integrated systems. The rest of this chapter sketches the different kinds of integrated systems that could be considered, in relation to the structures that are beginning to emerge in the electricity supply business.

7.3 Emerging electricity structures

Chapter 1 of this volume noted that the evolution of electricity systems in Europe is itself leading to important changes of scale. The vast integrated systems of national supply – generation, transmission, and regional monopoly companies dedicated to the distribution of that power – are being transformed by the progress of both technology and regulatory structures. The irony is that this combination is leading to changes of scale in both directions simultaneously. On the one hand there are the rise of technologies that can generate efficiently at small scales and of technologies that make it possible to track and allocate costs in local networks, and the liberalization that enables smaller, independent generators to access grids. The combination of these trends can encourage the rise of smaller systems in which local generation receives credit for generating close to the point of demand. As noted, this may be very valuable for renewables, which, like most electricity demands, are by their nature relatively small-scale and dispersed. This leads to the concept of distributed utilities sketched in Chapter 2, which ultimately could include local DC systems at the level of individual buildings or estates.

On the other hand, the European Commission's tenacious pursuit of electricity liberalization is aimed partly at enabling electricity to flow more freely across borders within Europe. This, combined with the steady improvement in long-distance transmission technologies (again DC), points towards greater international exchange of electricity, both for reasons of system management (e.g. spread of peak loads) and so as to increase the utilization of the cheapest sources of electricity in the region.

In reality there is no contradiction in these seemingly opposite trends, and it is the contention of this chapter that both trends are favourable towards greater utilization of renewable electricity. They are not contradictory because the trend towards localization does not, and indeed cannot, imply that all electricity needs can be met locally. It simply means that where some electricity can reasonably be generated towards meeting local needs, the full value of the avoided distribution costs and losses can be recognized. As we have seen in Chapter 1, this value may be considerable. Particularly for remote regions and some rural areas, the value is probably rising further because of the growing environmental and other constraints on further expansion and reinforcement of distribution systems. In these cases distribution costs and constraints may be more important than the costs of the bulk power supplied from the transmission network, so that lowering the cost of bulk transmitted electricity will not necessarily make local embedded generation redundant.

Indeed, in such cases local sources may economically meet a high proportion of the local electricity needs. In industrial and urban areas, however, with much higher density of demand and lower per-unit distribution costs, the fraction that can reasonably be met from renewables generation is likely to be much lower, except possibly in the special case of PV building façades (Chapter 5), and some of the energy from waste technologies considered in Volume III. All such systems are likely to draw, to some degree or another, on the traditional benefits associated with the diversity of large integrated systems: the ability to draw on external supplies, from cheaper production sources, when it is necessary and economic to do so. Localization increases local efficiency; internationalization similarly allows greater efficiency on a Europe-wide basis. We now consider how these trends relate to the use of primary renewable electricity sources and the cross-cutting policy issues that are raised.

7.4 Integrated local systems: a typology

As indicated in Chapter 5, PV has a whole set of expanding off-grid specialized uses, ranging from communication facilities to street lighting, as well as the supply of remote dwellings; the power is often supplied as DC to avoid conversion losses. On a slightly larger scale – the lowest level of what are generally considered 'electricity systems' – a wider set of renewable energy options begins to emerge. The 230 V networks that supply domestic dwellings and some light industry can meet total loads up to about 300–1000 kW, equivalent to a few hundred houses. DC systems could also be considered at this level, instantly improving the economics of PV and wind energy in particular, but this would represent a far greater change from existing systems and is as yet only a theoretical consideration for most supply systems; it also still too small readily to accommodate the most economically sized wind or hydro systems in many cases. It is at the next levels up, the 11 kV distribution systems that can accommodate 0.5–20 MW of power, and the 33 kV systems that may reach up towards 50 MW, that the greatest opportunities may arise. The appropriate combination of renewable sources at a local level depends upon local conditions. However, four main classes of possibilities can be identified.

Islands

As noted in Volume I of this series (pp. 42–7), islands (excluding the main islands of the UK and Ireland) account for about 4% of the European Union's population. Island supplies represent the best-explored options for renewables. Several studies of integrated packages have been conducted, and some implemented. Wind energy is the most common element. For the more northerly and westerly islands in Europe, this may be combined with small hydro power and coastal wave energy; all of these would tend to be correlated with the strong seasonal demand for winter heating and indoor lighting. In a few locations, tidal power might provide a more dependable input across the seasons. For isolated islands, the system demands can be met reliably by combining the intermittent renewables with diesel generation (e.g. on the Shetland Isles) and/or load management technologies such as interruptible refrigeration supply (e.g. on Fair Isle). For the long term this may well be more attractive than developing grid connections. Even for islands already connected to mainland grids, the costs and losses associated with sub-sea transmission to relatively small loads may make an integrated renewables supply attractive, perhaps ultimately leading to retirement of the link in some cases.

For Mediterranean and other southern islands, the seasonal demand profile may be weaker, or reversed to give summer peaks (e.g. because of tourism and air-conditioning). In these cases solar PV may assume much greater importance, often in combination with wind energy.

Coastal regions

For coastal regions, the options are similar to those for islands, but the scale may be larger and the competing cost much lower than for isolated islands because most such regions are likely to be grid-connected. Set against this, larger demand, a stronger network and the ability to sell power back to the system may enable more economic sizes for wind turbines particularly; easier access, compared with islands, will also reduce maintenance costs.

There are strong winds along most of Europe's west coasts and this may be the most obvious element for coastal regions, but small-scale tidal and wave input, harnessing the focusing characteristics of coastlines and perhaps incorporated into existing harbour and coastal defence infrastructure, may be more promising than for islands. The rain from the prevailing westerlies also adds hydro power to the mix of promising resources. The lands of the Celts in western Europe, stretching from Galicia in Spain, through Brittany and Ireland, and up to the Scottish Western Isles, thus offer particular and promising characteristics for the development of localized renewable energy systems.

Mountain village regions

Mountainous and upland regions have about 7.5% of the Union's population. Some mountain regions are still not connected to grid supplies; those that are involve long lines, with relatively high maintenance costs and transmission losses, and high environmental impact.

Supply options to upland villages can harness a mix of small-scale hydro, wind and PV. Both hydro and wind are likely to have good seasonal correlation, and to some extent short-term correlation, with demand (except for extreme upland dwellings that are abandoned in winter, for which PV is more attractive). Few such regions are connected to gas grids; other fossil fuels and/or biomass can offer heating but this might well be supplemented by electricity. With these characteristics, a combination of local renewables with heat storage, more efficient equipment and/or load management and electricity storage could enable much demand to be met locally and greatly cut down on peak winter loads, avoiding the need for grid connection or strengthening. Environmental pressures to avoid new cables, or even to strengthen existing ones, tend to be particularly strong in mountain regions, and may further increase the attractiveness of off-grid options.

These factors combined suggest that mountain villages in particular might benefit from attention in considering options for the development of small integrated renewable energy systems.

Rural distribution

The biggest category of 'local system' in most of Europe, excluding urban and industrial areas, is for rural electricity supply. As noted in Chapter 1, the demand is substantial and the electrification of rural areas is neither complete nor cheap in Europe; as illustrated by the data on France, in which rural demand has risen (despite the decline in population) to account for over 15% of total demand, expenditure on rural distribution reinforcement alone accounts for about one billion ecus a year, and the per-unit cost of that supply is estimated to be more than twice that of supplying urban centres.[1]

The local renewable energy options, however, are more limited than for the other areas discussed above. Winds are attenuated by the trees and hedges typical of rural areas. In some cases there could be opportunity for run-of-river hydro. The surfaces of agricultural buildings may offer natural locations for PV. For the most part, however, the biggest renewable resources are, not surprisingly, agricultural wastes and crops of potential biomass, considered in Volume III.

7.5 Integrated local systems: institutional obstacles and policy strategies

Given the technological developments sketched in the chapters of this volume (and for biomass-based electricity sources set out in Volume III), the potential to incorporate more integrated renewable energy contributions in local systems appears considerable. Why, then, has so little been implemented, and what policy and institutional developments might change this?

---

[1] Christophe de Gouvello and Marcello Poppe, *Maîtrise de la Demande d'Electricité et Surcoûts de la Desserte Electrique Rurale*, CNRS No. 940 (Paris: CIRED, 1994).

Four main reasons seem to explain why so little has been developed.

(1) The pace of technical progress itself. The developments that have made renewables more attractive and accessible are relatively recent, as are the developments in relevant power conversion and control technologies. There are few engineers, few companies and developers, and even fewer people among the potential purchasers of such systems, who are familiar with the possibilities.

(2) Industrial factors. The renewable energy industries are, for the most part, separated by technology, and approach sales from the point of view of promoting their technologies rather than meeting integrated needs of consumers and systems. Given the extent to which the potential of renewable energy systems may reside in integrated applications, this fragmentation of the supplier industries, generally into very small companies, is an additional obstacle.

(3) The institutional structure and cultural characteristics of the electricity supply industry. As sketched in Chapter 1, the industry has almost since its origins focused upon the goal of extending the scale of supply systems, generally with power generated from ever-larger centralized generating stations. Distribution is an extension both of the scale and importance of the company, and of the customer base for recovering the huge generation investments of the 1950s to 1980s. The extension of power grids has also been widely seen as synonymous with industrial development. Frequently, electricity distribution companies have been structured and mandated on the assumption that they are there to distribute power that is generated in central facilities, and their success has been measured in the reach of their networks and the amount of power transmitted through them. Many distribution companies are not in the generating business at all and may be legally barred from it. Thus there has been a huge institutional and cultural disposition towards extension and reinforcement of electricity distribution from central power stations, with no countervailing institutional structures that could promote local generation in a way that can be offset against distribution costs, even where this would be cheaper.

(4) A direct economic corollary of this institutional background is the structure of tariffs and cross-subsidies within the electricity supply industries, which supports grid extension and reinforcement and undermines any alternatives. This occurs at two levels. The French studies cited above have shown that larger rural consumers in France are systematically undercharged compared with the costs they impose on the system, and for southern Corsica they conclude that 'each kilowatt sold at 0.6 FFr/kWh (equal to 9.2 ecucents) contributes to the regional deficit of EDF'.[2] A more generic problem is that utilities generally try to set standard tariffs according to the volume of sales, independently of location. This implicitly subsidizes rural and remote supplies at the expense of low-cost urban consumers, in ways that are rarely transparent and which therefore tend to hide the full economic value of localized regional supplies.

The general underpricing of electricity to some classes of consumers reflects in part the interest of integrated monopolistic companies (especially when nationalized) in extending size and sales above profit. The use of uniform tariffs in part reflects technical constraints – the administrative burden of differentiated costs and billing – which are, however, being removed by newer information technologies. Far more difficult is the fact that uniform tariffs are perceived to be more equitable: in particular, charging higher tariffs to rural and remote consumers would be politically highly sensitive.

What institutional and policy responses could help to remove these obstacles? The first two have to be addressed primarily by the renewable energy industries themselves. Along with efforts to inform about the improvements in and potential benefits of particular technologies, the fragmentation of the supply industries points to the need for greater collaboration among the different technology-based companies, especially across technologies, and for greater efforts to be directed at meeting the potential local-market interests with such integrated systems.

---

[2] Christophe de Gouvello, *Maîtrise de la Demande d'Electricité (MDE) en Zones Rurales,* Rapport de Synthèse pour le Ministère de l'Industrie et ADAME (Paris: CIRED, 1996). The short-run marginal cost of supply in the department of Corse du Sud is estimated as 0.7 FFr/ kWh (10.7 ecucents) and the total cost as 1.15 FFr/kWh (17.6 ecucents).

The problems associated with the institutional and tariff structures of electricity supply in most of Europe demand more drastic remedies. But here, the 'drastic remedy' could be at hand with the liberalization of electricity systems, which (as outlined in Chapter 1) is beginning to emerge across Europe after more than a decade of political struggle. The essential need for renewables is that the changes be implemented in ways that reflect the full benefits of decentralized generation, as discussed in Chapter 2, and the full environmental and other external benefits that also arise from renewable energy. The resulting policy issues are explored further in Chapter 8 of this report.

7.6 Integrated large-scale systems for remote renewables: transmission options

We now turn to the issues associated with the integration of renewables at the very large end of the spectrum: pan-European systems for exploiting the greater wealth of primary renewable electricity sources in the sparsely populated regions of Europe's periphery.

Clearly, such resources carry no benefits in terms of 'embedded' generation; quite the converse, they would carry the additional cost of amalgamation and stepping up to a high-voltage DC cable, with transmission and delivery into Europe as a centralized source of power. They would thus have to compete directly against established centralized power generation sources, with the additional costs of long-distance transmission. However, as noted in Chapter 2, the costs of long-distance transmission form only a small fraction of existing system costs and have fallen considerably, to the point where Norway is building a series of sub-sea cables to link its hydro capacity more than 500 km across the North Sea to the Dutch and German systems.

The idea of using long-distance transmission to tap remote renewable energy sources is not new. Indeed, the most famous and prolific inventor of the postwar world, Buckminster Fuller, is credited with the grand idea of establishing a global interconnected grid, which would transmit electricity instantly to even out the fluctuating demands for power in different parts of the world. He argued that such a global system would help to unite nations and overcome regional tensions, and that electricity could even become a stable and exchangeable currency. He took it as a matter of course that most of the power would ultimately come from solar energy, with the grid removing the problems of variability and locality.

Many years later an organization was established to promote and explore this vision, attracting interest from engineers in many countries,[3] including extensive technical evaluations by Russian engineers of connections spanning Europe, Asia and North America.[4] In 1992 the US Institute of Electrical and Electronic Engineers held a special meeting to explore proposals spanning the transmission of Canadian hydro power and southern US solar power across America, transmission of hydro power from the Zairian Grand Inga to Europe, and transmission across the Middle East; a range of other proposals was covered in less detail.[5]

The real difficulty facing such schemes is not the unit cost of power, which in some cases is clearly competitive on a paper evaluation.[6] The problem is the sheer scale of up-front investment and risks involved. In theory the most ambitious Grand Inga project could deliver almost 50 TWh/yr to Europe at competitive rates, but it would require up-front investment estimated at US\$60 billion (1991 value[7]), originating in and passing through some of the politically and institutionally least stable countries. In reality, no bank would touch such an investment starting from scratch. The policy-relevant question is not whether such grand schemes, or a global grid, are desirable or even economic in principle. It is whether slightly more manageable and plausible projects might exist for Europe, which might perhaps evolve into wider integrated systems, for example with interconnection

---

[3] Global Energy Network International, PO Box 81565, San Diego, CA 92138, USA. GENI publishes a six-monthly newsletter carrying information about the development of proposals for very-long-distance interconnections.

[4] Yuri Rudenko and V. Yershevich, 'Is it possible and expedient to create a global energy network?', *International Journal of Global Energy Issues*, Vol. 3, No. 3 (UNESCO, 1991).

[5] 'Remote renewable energy resources: long-distance high-voltage interconnections', *IEEE Power Engineering Review*, Vol. 12, No. 6 (June 1992).

[6] The most ambitious Grand Inga scheme was estimated to deliver power to Europe at under two US cents per unit at 1984 prices and 5% discount rate, which would make it competitive even in terms of today's prices and higher discount rates.

[7] US\$60 billion (1991 value) is equal to about 55 billion 1995 ecus, when corrected using the consumer price index.

funded through the European Investment Bank in the context of developing trans-European networks.

The other possibility for conveying remote renewable energy would be to convert it to hydrogen for long-distance pipeline transport. Hydrogen energy has many apparent physical and environmental attractions, and for this reason has many proponents especially among engineers. However, as a means of transporting renewable energy it would have to overcome economic obstacles much more formidable than those facing long-distance electrical transmission. Hydrogen energy transport and use is discussed more broadly in Volume III; here, the focus of the discussion is on regional DC electricity cables.

Central and eastern Europe do not seem promising locations for generation and export of renewable electricity, with the possible exception of some large hydro schemes. Generation to meet local needs and reduce national dependence upon Russian energy is in some cases a possibility and, as noted in Volume I, Chapter 5, EU and other foreign assistance could have a large role to play in this. But the limited resources, the large domestic demand, and the political and economic conditions sketched in Volume I, Chapter 5, mean that renewable electricity is likely to play only a modest role, and it seems unlikely that renewable resources would be developed for regional trade and export, with the possible exception of some large hydro schemes.

Two other regions do, however, lend themselves to consideration. Europe's southern borders face the solar (and in some cases wind) resources of Mediterranean, Sahara and Arabian desert regions. The northern and western frontiers of Europe face large resources of sea-based renewable energy, and also the hydro resources of Norway and Iceland, with the latter's geothermal resources in addition. We consider each in turn.

7.7 Integrated large-scale systems: the Mediterranean and north Africa

Southern Europe's frontier regions have already attracted some interest in their potential for solar power generation, particularly solar thermal power. Chapter 5 (Box 5.1) argued that solar thermal power would remain economically unattractive for generation within Europe (with rare exceptions).

As a desert-based centralized power source, its situation is very different. The high intensity and reliability of the solar resource overcomes many of the disadvantages of solar thermal technology in European locations, and the modular nature of PV is no advantage in more remote, desert regions; indeed, it is a drawback compared with the intrinsic large scale of solar thermal power generation. In places, such power could be well supplemented by wind energy, for example the strong winds in some areas of the north African coast, in the Atlas mountains not far to its south and in some of the eastern Mediterranean.

Solar and wind-based power generation to meet the electricity needs of the Maghreb countries is a technical possibility. If it could be made technically and economically successful, it could play an important role in the region's development. Power generation is an important part of this development. A study of the region notes that energy consumption remains 'well below what would normally be expected [given the region's per-capita GDP] ... energy can become a very serious constraining factor to attain the objective of sustainable growth in the region'.[8] Furthermore, the study concludes:

A coordinated effort by all relevant institutions is required. It is almost impossible to finance all required investment purely on a commercial basis, without a quota of concessional finance from local governments and external funds. The most critical area is electricity power generation, transmission and distribution ...

Technically, solar and wind resources are large enough to make a major contribution to domestic generation and still extend exports to meet the still-growing power needs of southern Europe. Links from Morocco to Spain, from Tunisia to Corsica and Sardinia or Sicily, and from Egypt to Crete all offer themselves for consideration (see Figure 7.1). In the latter cases, the investments could take advantage of the relatively high cost and value of power in these islands, and could then link through existing cables to the European mainland.

---

[8] Franco Reviglio and Giacomo Luciani, 'Energy in the Middle East and North Africa: a constraint to economic and social development?' (Milan: Fondazione ENI Enrico Mattei, 1996).

Figure 7.1 Transmission options for the Mediterranean

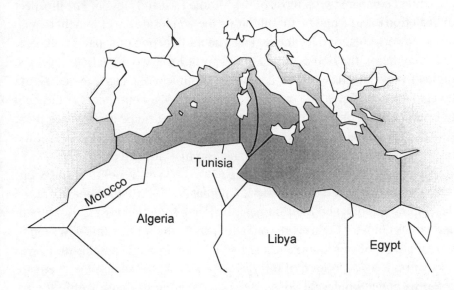

Source: Author.

One detailed technical study has been conducted of the possibility for solar exports from the Sahara.[9] This focused on grand schemes with the energy exported as hydrogen, perhaps because of the greater distance. As noted, this seems an economically implausible option for some decades. More credible is the increased evolution of solar and wind electricity production, probably in tandem with gas power generation. As indicated in Chapter 5, some of the most spectacular commercial developments of solar thermal technology have involved hybrid solar–gas systems, rendered economic by joint use of the boiler and power conversion equipment. Given the conditions in the Maghreb, this must be an interesting option, particularly for Morocco and Tunisia, which rely on gas imports from their potentially unstable neighbours.

---

[9] J. Ogden, in T.B. Johansson et al. (eds), *Renewable Energy* (Washington DC: Island Press, 1993).

Furthermore, although the gas resources of Algeria are considerable, they are limited compared with those of the Middle East and Russia. The prospect of depletion may be one factor inhibiting the wholesale development of gas-based power generation for the region. The introduction of solar–gas generation, together with wind energy where appropriate, would help to meet the region's power needs while extending the lifetime of the gas reserves. At the level of specific fields and projects, a growing solar component could help to extend the lifetime and ultimate profitability of fields as their gas pressure declines, thus starting to realize at a project – and regional – level the broad conceptual idea of gas as a 'transition fuel' towards a solar age.

What makes this possibility of particular interest is the political and economic context sketched in Volume I, Chapter 5. The Mediterranean area is developing rapidly, and the European Union is keen to enhance investment and growth in the Maghreb region, partly in the belief that this will help to stem the flow of migration and the rise of Islamic fundamentalism. Power generation is a crucial part of this development. At the same time, there are concerns about wholesale dependence on Algerian gas both within the region and in southern Europe, because of worries about the political stability and possible terrorist disruption. The fact that the leading companies in solar thermal power generation are German and Israeli might also assist political support in the Union, perhaps in the context of the Middle East peace process. Promoting such development would appear to accord well with the Union's political, economic and environmental goals.

The economics of such developments, including the relationship to possible power exports to southern Europe, need much further study, and the technology of hybrid solar–gas generation clearly requires further technological and industrial development. But given the progress made, the prime need is to start building industrial capabilities and experience so as to hone the technologies and bring the price down. An important institutional obstacle at present is the government control of power systems in the region, with the concomitant focus on large, centralized power generation at subsidized prices. However, Morocco is already experimenting with efforts to open up the power sector and attract foreign investment. Without any government role, this is no more likely to lead to the development of solar thermal power than at present, but given the context – and given the technological strides made in the Californian investments – EU and local governmental support for private investment in solar thermal power generation could yield rapid progress.

To follow on the THERMIE demonstration project on solar thermal power in Morocco, it would thus seem appropriate to explore the scope for some of the ongoing European Investment Bank investment to be directed towards supporting solar (as well as wind) energy developments. It is too early to say whether this really might ultimately lead to renewable electricity exports to southern Europe – but the possibility cannot be ruled out, and the path in itself seems potentially valuable. As with wind energy, European industry and the Mediterranean region may be able to capitalize on the Californian experience, and reap multiple benefits.

7.8 Integrated large-scale systems: northern Europe and the North Sea

On the other side of Europe renewable resources are also abundant, but very different in character. Norway and Iceland are both sizeable, mountainous and sparsely populated countries with correspondingly large hydro power resources and potential surplus (though environmental opposition to new schemes in Norway limits the likely surplus, and gas turbines are being considered to produce power for export). In Iceland, hydro power is joined by geothermal energy which is potentially available in very large quantities; geothermal energy production has also been proposed from depleted oil wells in the North Sea (see Volume III).

In addition, the seas themselves contain large renewable resources. As noted in Chapter 4, the stronger winds offshore can help to offset the additional costs of offshore wind turbine operation, and represent a huge theoretical resource, both more intense and more consistent than those onshore. The main obstacles are the costs of the foundations and/or moorings, of maintenance and of transmission to shore. Transmission costs are also an important part of wave energy schemes; the UK government assessments of 2,000 MW wave energy devices generally estimate transmission to account for about £500 million (as well as power losses of 5–7%), over 10% of total capital cost for the cheaper devices.[10]

---

[10] T.W. Thorpe, *Wave Energy Review* (Harwell: ETSU, 1992); £500 million is comparable to 800 million 1995 ecus.

Figure 7.2 Sub-sea transmission options for northern and western Europe

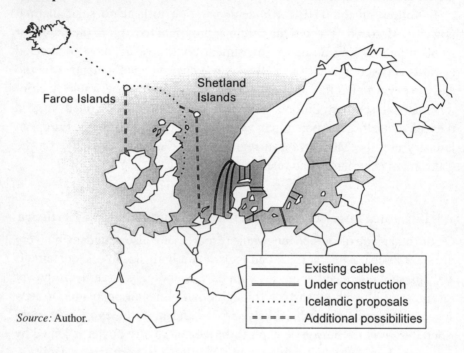

Source: Author.

Transmission developments

One factor which may alter the context for exploiting some of this renewable energy is developments in long-distance transmission – both the technology and actual projects. Seven high-voltage DC cables link Norway and Sweden to Denmark, Germany and Finland. In the early 1990s, the Norwegian Power Grid Company Statnett signed agreements for three new sub-sea cables to link Norwegian hydro capacity to western Europe: two to land in Germany, one in the Netherlands. Each cable will run directly through the North Sea, be about 540 km in length and transmit 600–800 MW at a voltage of 450–550 kV DC.[11] Various transmission options are displayed in Figure 7.2.

---

[11] Statnett, *Annual Reports* (1994, 1995); Information releases (Oslo: Statnett, 1995).

Figure 7.3 Unit cost of Icelandic power exports as a function of transmission distance

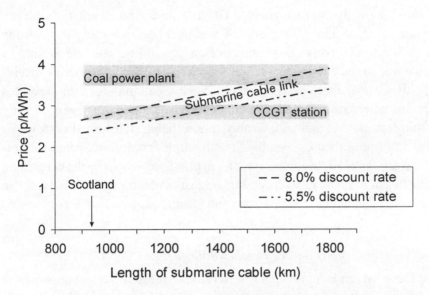

Source: E. Guðnason et al., *3rd International Conference on Power Cables and Accessories 10 kV to 500 kV* (London: November 1993).

Iceland has for some years advanced ambitious plans to lay a cable to northwestern Europe.[12] The initial proposal focused on Scotland, with alternative routes to Norway or to Northern Ireland. The initial proposal was for a 1000 km line of around 500 MW carrying about 4 TWh/yr, but the practical resource of hydro plus geothermal energy is estimated to be ten times this. The unit cost of the power is displayed in Figure 7.3. The main obstacle has been the lack of interest in the UK, given that Scotland already has surplus generation, and onshore exports further south are constrained by the existing transmission capacity together with environmental and institutional objections to creating additional north–south transmission capacity in

---

[12] *400 kV HVDC Submarine Cable Link Iceland–Faeroe Islands–Scotland* (Landsvirkjun: National Power Company of Iceland, 1988).

the UK. Subsequent proposals included linking the line on to East Anglia, but with the recent 'dash for gas', the UK does indeed have an overall surplus.

With the steady improvements in DC transmission technology, however, the option of circumventing the UK's abundance and going through the North Sea to Germany could emerge as a possibility. The relationship of such a line to the Norwegian cables would have to be addressed: conceivably, they could meet in the northern North Sea, and share an expanded capacity. Alternatively, the Icelandic exports could be routed west, landing in Northern Ireland and conceivably thence being transmitted on towards France or (more likely, given the French surplus) northern Spain or Portugal. An attraction of this western route, in principle, would be the continuing high rates of demand growth and higher cost of electricity in Ireland, Spain and Portugal, and less competition from cheap fossil-fuel sources.

Possible implications for ocean renewable energies

One factor which increases interest in these developments and possibilities is the opportunity for helping to open to exploit sea-based renewable energy. As noted, transmission costs are a major impediment to this exploitation. If, however, sub-sea cables were installed for other reasons, significant savings could arise from deploying renewable energy devices near them and using the existing DC cables to transmit the power onshore. This would, of course, require either that the cables are laid with spare capacity, or that the offshore generation displaces some of the generation at the source of the line. Neither seems implausible, particularly since both the Norwegian and Icelandic options involve considerable hydro power, effectively adding storage to the intermittent generation and yielding high overall utilization of the lines.

Perhaps more importantly, this would allow a more incremental approach to installing offshore renewable energy capacity without the huge up-front costs of a dedicated transmission cable.

Routings from Norway and Iceland would pass through regions of prime interest for ocean-based renewables. The North Sea is windy, and lines particularly to the Netherlands and Germany could pass over the Dogger

Bank area, which might offer seabed foundations for wind energy. In the central and northern North Sea, and moving out into the Atlantic, the wave intensity also increases. Lines from Iceland directed east of the UK would pass through regions of high wave intensity, and also through the regions of maximum tidal flow currents (Chapter 6). Lines directed from Iceland to Ireland – particularly if ultimately extended to the Iberian peninsula – would traverse the major North Atlantic wave energy resource.

Clearly there are different levels of plausibility and different timescales for these options. Although the proposals for Icelandic exports to the UK are technically quite well developed, there is no real sign of a consortium emerging to promote and invest in this or the still bigger schemes required to circumvent the UK's current surplus capacity. Furthermore, the prime renewable energy technologies involved – wave and perhaps tidal stream energy – are still mostly at the stage of desk or model designs of questionable economic viability. Nevertheless, the overall vision is intriguing and would seem to justify a much greater level of RTD support.

The situation regarding Norwegian exports and North Sea wind energy is rather different. Contracts for three cables, all due to enter service by 2003, are already signed. Wind energy is a commercial industry and experience of offshore operation – though not deep-sea siting – is accumulating, as outlined in Chapter 4. In addition, there is another factor that may encourage such developments – the existing North Sea conventional energy industry.

Relationship to North Sea oil and gas infrastructure

Since the early 1960s, over £100 billion has been invested in oil and gas in the UK Continental Shelf, most of it in the North Sea, with probably as much again from other countries. By 1994, more than 400 production and related platforms were installed. Most are still in active production, but many are due for retirement over the next 10–20 years as fields deplete (Figure 7.4). UK capital expenditure on new developments is falling during the 1990s and is projected to decline to low levels after 2000, being replaced in part by expenditure on decommissioning old facilities.[13]

---

[13] Arthur Andersen Petroleum Services, *Evaluating Innovative Ideas for Platform Re-use and Determining their Cost-effectiveness vs. the Disposal Option* (January 1995).

Figure 7.4 Remaining years' operation for UK oil and gas fields

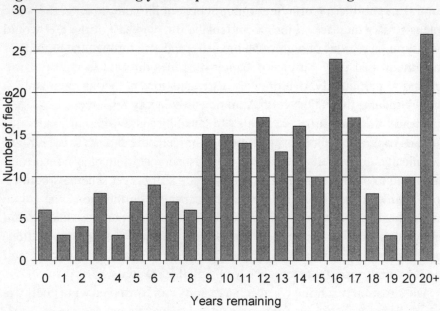

Source: Arthur Andersen (Note 13).

Although a handful of platforms have already successfully been decommissioned, the furore over the attempted disposal of the Brent Spar oil storage buoy in 1995 ignited interest in both the environmental and the economic aspects of decommissioning old platforms. It is not cheap: a report in 1995 estimated the total costs of abandonment in the UK North Sea sector alone to be about £7 billion, with estimates running up to £10 billion in the industry (8.4 billion and 12.1 billion ecus respectively).[14] The legal framework, which is similar in the UK and Norway, prevents companies from simply walking away from the problem: they are obliged to incur the costs of dismantling unless an acceptable alternative use is found for the platforms.

---

[14] Ibid.: Michaal Bishop, 'Economics and policy for the abandonment of oil and gas platforms in the North Sea and an analysis of the potential of re-use for wind energy production'. Report to RIIA, March 1996.

This raises the question of whether abandoned platforms might provide the foundations and/or mooring for offshore wind (or wave or tidal stream) energy that would otherwise be so costly. The answer is not known. The platform would provide foundations and a base for maintenance, both extremely valuable for offshore technologies. In Chapter 4, the results of studies were summarized that suggest that the limiting factor for wind energy installed on abandoned platforms would be the cost of transmission, given the relatively small capacity that could be installed on any one platform (which would also limit the resource). Even if five 1 MW wind turbines could be installed on one platform, transmission costs would be prohibitive for distances beyond 10–20 km.

The passage of sub-sea cables through the central and southern North Sea – perhaps with development of a ring grid to collect wind power from a number of platforms in the same region – could possibly make the use of abandoned platforms in this way more viable. An alternative approach could emerge if floating platforms for wind energy – perhaps tethered to fixed oil or gas installations – were to prove viable. Some of the southern North Sea platforms are also in areas with relatively strong tidal currents.

In the central and northern North Sea, the existence of platforms might also affect the economics of wave energy, of which the cost of anchoring is an important component. This could be combined, of course, with wind energy (Chapter 6). Again the existence of transmission capacity and/or existing anchoring infrastructure would presumably help to improve the overall economics.

Prospects and policy implications

These visions should be treated with due caution and scepticism. With current technology, the North Sea options are unpromising and the Atlantic options are speculative in the extreme. The offshore environment is hazardous and costly. Yet history demonstrates the danger of dismissing technologies and systems that are young and plainly immature. The key point evident from the chapters on the technologies concerned (and indeed the history of North Sea oil and gas operations) is that new ideas keep emerging and estimated costs keep declining. At present it is not possible to make a realistic evaluation of these various possibilities, or to select the most promising. Not only are there too many fundamental engineering uncertainties about the costs involved. There are also too many different possible combinations of remote generation from Norway and Iceland, multi-purpose uses of associated transmission capacity, and links with existing or new oil and gas platforms, for any convincing conclusions to be reached as yet.

What is striking, rather, is the large potential resource and the apparent synergies of interest and timescales. A huge and innovative offshore industrial enterprise is approaching the end of its major capital investment phase. As well as needing to explore new horizons to minimize the costs of contraction, it is turning its attention towards the problem of decommissioning platforms. The first phase of decommissioning investment, beginning in earnest in the first decade of the next century, will focus on the older platforms in the shallower, calmer waters of the southern North Sea. At the same time, Norway will be laying cables through the region. Over the next ten years also, wind turbines could be installed much more widely on existing platforms, and the basic technologies for large-scale offshore wind energy generation in shallow waters will have gained some years' experience. A modest industry of offshore wind generation, in collaboration with the oil and gas industries, could be emerging.

Over the subsequent decade, attention could turn more to the central North Sea region. With greater depths and stronger waves, the engineering challenge is greater. Options for floating, tethered structures in place of platforms or seabed mountings would need exploration; and trial experiments to integrate wave energy into these structures could be undertaken.

Finally, probably into the third decade of the next century, a fully fledged and free-standing offshore renewable energy industry could begin to take shape. By then, Icelandic investments might have laid cables through the northern North Sea and/or western Atlantic seaboard, perhaps easing the transmission problem. And the limitations of gas supplies to Europe – maybe including CO_2 constraints – might begin to be clearer, justifying the higher investments that would probably be required.

It is by its nature hard to predict the contributions involved. However, wind energy is the most developed of the technologies, partially proven already for near-shore applications, and deployment could start in the earlier stages of such developments. Given the resource and the system assessments discussed in Chapter 4 and taking account of probable limitations on bulk transmission to north European markets, developments such as those sketched above might culminate in offshore wind energy supplying 1–6% of European electricity by 2030. Wave energy developments would be slower and more limited, with expansion mostly as this process moved towards the northern North Sea and Atlantic regions, perhaps reaching a contribution of 1–3% of European electricity by 2030 – still, by then, a huge industry. Tidal energy remains limited and speculative, with plausible contributions below 1%.

The point of this discussion, as indicated, is not to predict that such contributions will necessarily prove feasible and economic. Rather, it is to sketch the possible elements that would need to be brought together over time for such an industry to evolve, and hence point towards an incremental strategy that would need to be developed to explore it. At present, a strategy to realize the above contributions would need to comprise the following elements:

● continuing (perhaps dedicated) capital and/or output subsidies to encourage deployment of offshore wind energy schemes (see Chapter 4);
● modest support for the deployment of offshore wind energy to supply power on operating oil- and gas-production platforms, in return for feedback on the costs and performance, to gain engineering knowledge about the costs and characteristics of operation in such an environment and to improve the technology;
● continuing RTD efforts to improve offshore wave energy devices and to explore whether any credible approaches to tapping tidal stream energy exist;
● more focused studies, leading to prototype trials, of floating wind and wave energy devices, probably anchored to existing production platforms – perhaps ones already connected to mainland grids;
● re-evaluation of the economic studies of offshore wind and wave energy in the light of the knowledge gained and taking into account the possibility of connection to existing sub-sea transmission;
● incorporation of the above possibilities into the evaluation of decommissioning of North Sea platforms;

- cooperative studies with the relevant Norwegian and Icelandic interests, to examine whether the possibilities for such offshore renewable energy generation may hold implications for the routing, capacity or other design aspects of proposed sub-sea cables.

What is striking, at least at present, is how relatively little such preparatory steps would cost.

Chapter 8

Conclusions and policy implications

Simple unit-cost comparisons of primary renewable electricity sources are misleading, because the value of electricity from such sources may differ from conventional sources in a dozen different ways. In addition to the six possible 'external' benefits identified in Volume I, six internal 'systemic' issues can be identified, associated with the dispersed, small-scale, modular, time-dependent and expanding market characteristics of primary renewables. Very little of this is captured in standard economic appraisals. The additional value arising from these characteristics is in most cases (but not all) positive, and capturing these factors is a key to sustaining expanding renewable energy contributions. This requires greater differentiation of electricity as a product.

Capturing the localized benefits of distributed primary renewables requires greater localization of European electricity; this could be a natural consequence of liberalization, which could also encourage 'green pricing' for consumers willing to pay more for renewable electricity. Exploiting the high-intensity resources, and those depending on multiple uses of infrastructure (such as offshore renewables) also could benefit from liberalization that allows new entrants and unimpeded international flow of capital and of electricity. Localization and internationalization are not contradictory trends. Technically, they reflect the better matching of sources and loads: distributed generation used for meeting dispersed demands, and more concentrated resources exploited and transmitted for meeting more centralized municipal and industrial demands. Politically, it also accords with the ongoing 'hollowing out' of states in Europe in favour of both local and EU-level decision-making, as appropriate to the issues involved.

Expansion of primary renewables will, however, be sensitive to public policy towards the architecture of liberalization, the internalization of external costs in generation and distribution, and specific renewable promotional policies. However, as the contribution of renewable electricity grows, the current systems of support will come under increasing pressure. Probably the most appropriate longer-term approach would be a system of 'renewable portfolio standards', requiring supply companies to produce, or to obtain credits equivalent to, specified percentages of renewable energy contributions. The primary advantage of this approach, which has recently been suggested by the European Commission, is that it encourages highest-value and thus least-net-cost applications, unlike present support systems; it can also readily be made compatible between different countries. Introducing such a system for non-hydro renewables will have little impact on inter-firm competitiveness at present, but it will become politically more difficult the longer it is delayed.

Currently available modelling studies underestimate the likely contribution of primary renewables because they cannot capture the likely changes in technology, infrastructure and institutions that will accompany expansion of renewable electricity sources. The analysis of this book suggests that by 2030, the primary renewables together are likely to supply between a quarter and half of European electricity. The biggest contributions would come from hydro and wind energy (onshore and offshore), but these sources would by then be almost fully exploited; the contribution from indigenous PV, and from oceanic and desert-based resources, could continue growing thereafter.

Liberalization is a necessary precondition for larger contributions from renewables, but not a sufficient one: in terms of information and RTD, investment timescales and incorporation of external impacts, it is disadvantageous to renewables, especially in comparison with combined cycle gas turbines. Although CCGTs are technically, economically and environmentally highly complementary to renewables, over the next couple of decades investment in CCGTs could dominate to an extent which inhibits the development of all other sources including renewables. The prospects for primary renewables over coming decades will depend upon the readiness of governmental authorities – local, national and European – to recognize electricity liberalization as a means to, but not the end of, achieving strategic policy goals of efficiency, diversity and sustainability in European electricity supplies.

8.1 Introduction

The primary renewable electricity sources have developed greatly since the exploratory RTD of the 1970s. Except for large hydro schemes, however, their contribution to European electricity remains small; in 1995, under 3% of European electricity was derived from wind and small hydro combined. The contribution has grown rapidly in the 1990s – by 15–30%/yr – but this has been heavily dependent upon specific government policies for promoting renewables, through a mix of RTD and 'capacity-building' efforts led by the European Commission, and capital grants and market supports undertaken by many states in the Union.

The politics of Europe place limits on the direct expenditure that can be expected through European institutions. Continued growth in renewables ultimately depends upon investors making a profit, but the inexorable liberalization of electricity systems may also set limits on the extent to which governments can create special protected markets for renewables that depend upon transfer payments from other electricity producers or consumers.

So how sustainable is the recent growth? What is the real value of primary renewable energy, to what extent can investors capture this value, and what are the credible contributions on different timescales? This concluding chapter addresses these questions by bringing together the observations of the other chapters, and digging deeper into the structural issues surrounding European electricity.

8.2 The holistic value of primary renewables

A central thesis of this book, illustrated in varying ways in the individual chapters, is that the full value of renewables derives from a number of different characteristics, several of which are not reflected in the organization of current electricity systems. In addition to the direct unit value of electricity *per se*, additional sources of value can be characterized as follows.

First and foremost, the primary renewables share in common with most renewables several of the six external benefits discussed in Volume I:

(1) *Environmental benefits*, owing to the absence of emissions.
(2) *Increasing diversity* of energy supply in Europe.
(3) *Contributions to rural income* (less applicable to the primary renewables, with the occasional exception of wind energy).
(4) *Structural benefits*, owing to the concentration of resources in less developed regions of Europe.
(5) *Employment benefits*. As discussed in Volume I, claims about employment benefits need to be treated with caution (for this reason, in this volume we have avoided citing figures of jobs per unit of electricity for different sources), but the arguments may have some validity – for example, relating to PV in the construction industry, and ocean-based renewables as alternative sources of employment in declining offshore industries.
(6) *Technological benefits*. European companies have a strong international position with respect to most of the primary renewables, maintenance of which depends upon an expanding home base.

In addition to these benefits, however, there are other sources of value that are *internal* to the system, derived from the unique characteristics of the primary renewable electricity sources.

(1) *Flexibility advantage of small-scale, rapidly constructed sources*. With the exception of large hydro and some ocean-based technologies, all the primary renewables come in the form of relatively small, modular technologies that in many (though not all) cases can be relatively rapidly installed. With prospects for electricity demand and fuel prices uncertain, these characteristics have significant economic advantages. The rapidity with which such sources can be installed enables construction of large, long-lead-time and capital-intensive sources to be deferred (or avoided altogether, if demand growth proves to be lower than expected). Various approaches to evaluating such flexibility as an economic characteristic suggest that this represents significant benefit relative to traditional large-scale and centralized investments; one simplified study suggests a benefit of about 10% as compared with nuclear

investments.[1] The integrated, monopolistic electricity systems have been protected from such risks, but the value of small-scale and rapid construction will become more apparent in liberalized markets.

(2) *Risk and overhead advantage of modular manufactured sources.* A related but distinct feature is that most renewable technologies, once established, are manufactured products that can be purchased from manufacturers at fixed prices backed by performance guarantees; also, because they are capital-intensive, the cost is relatively certain, unlike gas turbines, for which the cost will depend upon potentially uncertain gas prices. This makes them less risky investments. Also, because they can be bought 'off the shelf', they may involve lower overheads for the generator than large centralized sources, which generally require extensive and ongoing utility involvement. Paradoxically, however, renewables may be charged with proportionately greater overheads if these are calculated as a fixed percentage of capital costs, as is common practice in utilities.[2]

(3) *Correlation with electricity demand.* Chapter 2 explained that the value of primary renewables depends upon the way in which the output varies over time, and demonstrated that such variability does not necessary penalize renewables, but rather that the value depends upon the relationship between source and demand variations. For contributions up to 5–10% of system electricity from a given source, statistical independence – an absence of any identifiable correlation – means that the value of electricity (other things being equal) is comparable with that from conventional baseload plant; a positive correlation makes it more valuable, and vice versa. As discussed in Chapter 7, Section 2, the output from most primary renewables in Europe is positively correlated with the relevant electricity demands and so – contrary to popular wisdom –

---

[1] 'If a short-lead-time option with an expected cost of 110% of the base cost of nuclear power is available, it is comparable in expected cost terms', Chris Chapman and Stephen Ward, 'Valuing the flexibility of alternative sources of power generation', *Energy Policy*, Vol. 24, No. 2 (February 1996).

[2] Shimon Awerbuch, 'Capital budgeting, technological innovation and the emerging competitive environment of the electric power industry', *Energy Policy*, Vol. 24, No. 2 (February 1996).

their variability in most cases adds to the average unit value of their output, rather than detracting from it.

(4) *Value of embedded generation.* These effects could be amplified by the potential of embedded generation in the network, i.e. reduction of line losses and distribution investments being required if local sources can supply local demands. As discussed in Chapter 2, 'embedded' benefits are complex to evaluate and highly site-dependent, but particularly for locations at weak points in networks, for remote locations or for very small-scale generation close to the point of demand (e.g. PV on buildings), they may be substantial.

(5) *Synergies with other 'clean' energy sources.* As discussed in Chapter 7, there are a number of possible synergies between 'clean' energy sources. Combinations of wind, hydro, wave and solar sources produce more 'firm' output than any one of these sources in isolation, and they interact well with storage. There are also important synergies with the characteristics of natural-gas power generation. Technically, this is most obvious with respect to the role of gas turbines as flexible, fuel-intensive 'load-following' plants, a natural complement to the baseload but variable renewables.[3] Another economic synergy arises because the cost of gas power generation is dominated by the fuel cost, which is uncertain: as generating companies become more dependent upon gas they will become more vulnerable to price shocks or interruptions. Indeed, since both gas and coal prices are sometimes related to oil prices, all fossil fuels may involve related risks. Renewable energy could help to reduce corporate risk by diversifying the fuel mix with sources whose generation cost is independent of such factors.

(6) *Technological learning and experience.* A final potential benefit associated with investments in the newer primary renewable sources may be the learning and economy-of-scale effects. Because installed capacities are currently very small, an additional 100 MW, for example, may represent a significant market expansion, which is generally recognized

---

[3] This may appear an unusual use of the term 'baseload', which is normally associated with low-fuel-cost conventional plant with constant output, but it is used in its correct technical sense as signifying plant which has very low avoidable cost and hence is used whenever available.

to yield cost reductions. In some respects this is a classical 'learning-by-doing' external benefit, but it may also be quite significant in terms of the experience it may give to the investor; this is a strategic benefit that is inherently difficult to quantify.

Together these comprise six systemic benefits that may be associated with many primary renewable energy investments, and that complement the half-dozen potential *external* benefits identified in Volume I, and summarized above. To these might be added a less tangible benefit that, according to innumerable surveys, people prefer renewable energy. Studies differ as to whether and how much people are willing to pay for that preference, but the general sentiment is clear and strong.

Almost all these issues contrast with the traditional means of analysing electricity economics, which is to assume that any one unit of electricity is like any other, and hence that it is the cost per unit which matters. Fundamentally, this analysis points to the importance of *differentiation* of electricity as a product. The real value of one unit of electricity may be very different from another, depending upon all the above issues: its impact on local and global environments and the diversity of the system; the time taken to install the generating plant, riskiness and location in the system; its correlation with demand; its relationship to other sources on the system; and so on.

Very little of this is captured in standard economic appraisals. Consequently, simple unit-cost comparisons may be a very poor guide to the full comparative value of different sources. The traditional methods of appraising electricity investments favour traditional sources, and help to justify their dominance of the system.[4] In turn, the methodology is justified by the institutional structure of a monolithic, centralized system in which neither the customer nor the system itself has information about the nature of the different kilowatt-hours being produced. And it is this structure, among other things, that has begun changing in the 1990s, and that in turn will start to change the realized economics of renewables.

---

[4] A case argued most convincingly by Awerbuch, 'Capital budgeting', op. cit. (Note 2).

8.3 Niche markets and the localization of European electricity

A major theme of this book has been the role of niche markets for renewable sources, such as:

- isolated dwellings (particularly for PV),
- islands,
- mountainous areas,
- coastal areas,
- offshore power demand (e.g. rigs and gas pumping),
- buildings, especially service-sector,
- weak points in distribution systems and
- rural supply.

In each of these applications, local electricity generation may have a higher value than power transmitted directly into the bulk system. Table 8.1 illustrates some of the different potential niche markets and the renewables that could most appropriately supply them, with a rough indication of the potential value. Recognizing such value in the prices received is important in developing sustainable markets in which renewable energy could compete unaided, which is important in its own right and as a stage in developing viable and self-sustaining renewable energy industries.

At present the benefits of niche markets are rarely reflected in the price paid. A century of centralized development of electricity systems, often as a state or regional monopoly, has led almost universally to the philosophy that tariffs should be uniform across consumer groups of a given size. As discussed in Chapter 7, this has been a matter of both organizational simplicity and equity. Indeed, in many countries, especially but not exclusively in the developing world, not only is pricing uniform across consumer groups, but for many groups it is heavily subsidized.

Since it makes little sense to pay independent generators more for electricity production than consumers pay for its consumption, in these circumstances independent generators (even where legal) rarely receive the full value of niche generation. At the same time, the centralized utility has had little incentive to undertake the laborious task of calculating the value of embedded generation at different points in the network and investing in

Table 8.1 Niche markets and applications for primary renewable electricity sources

Application	Share of EU electricity consumption (%)	Competing costs (∈/kWh)	Relevant renewables[a]
By location			
Off-grid houses, holiday homes	< 1	10–50	Small PV/wind-battery systems
Islands	2–4	5–20	Wind, small hydro, biomass, shoreline wave, PV
Offshore operations	1	5–20	Wind, wave
Mountain regions	4–7	5–20	Wind, micro-hydro, PV, biomass,
Rural grids	10–15	4–8	Agricultural wastes, biomass crops, wind
Municipalities	30–50	3–5	Municipal waste, PV-buildings
By end-use sector			
Buildings	60		
Service		4–10	Biomass/waste cogeneration, PV cladding and rooftops
Domestic		5–15	PV-cladding (DC)
Transport			
Marine	< 1	10–20	Onboard PV-battery
Rail	2	2.5–4	PV embankments
Electric vehicles	< 1 (at present)	?	PV embankments, garages, car parks
Industry/centralized	All grid	2.5–3.5	Large hydro, strong/offshore-wind, good tidal, wave, imports (hydro, solar thermal, geothermal)

[a] Waste and biomass here includes electricity generation from all forms of wastes and biomass; these options are considered in detail in Volume III, as is geothermal energy.
Source: Author.

numerous dispersed sources: it has been far simpler to follow the philosophy of expanding scale and invest in large, centralized plant. The same structures also shielded utilities from worry about the investment risk of such large, inflexible plant, since government ultimately underwrote all the risks.

Until at least the late 1970s, given rapid load growth and the poor development of small-scale sources, such structures probably made reasonable economic sense, but in the world of the 1990s it does so no longer.

To realize the potential systemic benefits of primary renewables set out in Section 8.1, the concept of distributed resource systems discussed in Chapter 2 needs to be implemented in practice. This raises new regulatory issues about operation, ownership, control and transparency of costs that will need to be resolved. At least three approaches can be identified.

One is to conclude that the complexities of 'distributed' operation can only be managed through centralized and integrated utilities, with highly developed computer-based analyses of systems and options, to determine appropriate utility-led distributed investments.

As discussed in Chapter 1, however, the trend in much of Europe is towards competition in generation markets. In some systems this involves bidding by independent generators for the right to install new generation to serve the system. The details of these emerging generation bidding systems are not important here, but the concept of multi-attribute evaluation of generation project bids is. Many evolving bid systems are based on the central-station utility paradigm, and value only the capacity and energy contributed by the bid project to the system. Some regulators, such as those in California, are looking at valuing other attributes (e.g. environmental impact), but the central-station focus remains. If distributed generation options such as PV are to become significant, then in such systems the regulators must either restrict bidding to only meeting the utility's true needs of central-station generation (and let the utility install generation cost-justified 'at the sub-station') or support the emergence of multi-attribute bidding systems that properly value location- and time-specific attributes.

The third approach – in sharp contrast to the first – is to argue that the complexities of distributed operation *require* a fully liberalized system, in which prices are developed to reflect the real costs at different times and places in the network, and the value of different power-source characteristics. Private generators then make their own choices about the kinds of equipment to install. This – pushing the process of liberalization down into the network, with differentiated and cost-reflective pricing – is in the author's view a plausible extension of the current moves towards liberalization, and in fact the one most likely to deliver the benefits of distributed resource systems in the long run.

As noted in Chapter 1, liberalization can take a number of different forms. The heavily regulated, rule-driven 'Anglo-Saxon' model would establish a truly competitive market with transparent rules, but it could become extremely complex to administer and enforce differential pricing over time and space, and explicit regulated incentives would be required to reflect external benefits of environment, diversity etc. The 'Continental' models, some of which are hybrids between the second and third options noted above, could give a greater discretionary role to governments in determining the kind of independent renewable energy projects that utilities would be expected to accept, and upon what terms. This would give greater scope for governments to encourage renewables in less formal ways; an understanding with utilities on such terms might emerge from the current vociferous debates over the compulsory buy-back tariffs in Germany and Denmark. In systems such as that likely to emerge in Italy, with competitive generating companies selling to a 'single-buyer' transmission company, the distribution companies themselves may have the strongest motivation to develop dispersed renewable sources.

The policy issues extend beyond the question of whether and what kind of decentralized, market-driven systems are better at capturing the economic benefits, however. There is another important policy dimension. Decisions on local investments involve far more than just economic questions. The introduction of more renewables will have some impact on the landscape, and they may interact with other local resources (such as pastures for wind, or farm buildings for PV). Knowledge and judgments about appropriate deployments in these circumstances often lie best with the local community. A centralized utility could never take an interest in such details for an energy input which is very minor in comparison with the total system. Hence, tapping the full potential for 'embedded' renewables depends upon local decision-making, which can be achieved only through liberalization of the electricity system.

Such 'localization' of electricity systems would not increase the value of renewables in all locations, of course. It would, for example, make the economics of using wind energy to supply city electricity demand worse, not better. The point is rather that the structural change arising from liberalization is towards greater *differentiation* of electricity as a product, and as argued in the previous section this is what is required to develop more sustained markets for renewables. Electricity will become less of a commodity, and more of a varied resource, tapped when, where and how it best matches the varied needs and preferences of the users.[5]

8.4 Concentrated resources and the internationalization of European electricity

At the other end of the spectrum are the areas of concentrated renewables that in principle could be collected on a large scale and transported long distances. The prominent possibilities are Norwegian hydro power, wind and wave in the North Sea and Atlantic coast, some tidal energy sites, Icelandic hydro and geothermal energy, and solar and perhaps some wind energy in north Africa. At present, only Norwegian hydro and the Icelandic resources are remotely economic, for the technology for offshore energies and even for solar thermal power is (as discussed) still relatively undeveloped. The conditions required for the development of each differ, but the common themes are *integration* and *international access*.

As discussed in Chapter 7, the exploitation of North Sea resources would appear to hinge not only on major technological developments but also upon development of multiple uses for the same infrastructure: wind, wave and/ or tidal current energies, drawing upon facilities that may have been developed originally for other purposes, such as oil and gas platforms, shipyards, and/ or sub-sea power cables from Norway or elsewhere. In north Africa, solar

---

[5] One recent study explores this issue in detail, focusing upon consumer attitudes, learning curves and resource characteristics: see Gary Nakarado, 'A marketing orientation is the key to a sustainable energy future', *Energy Policy*, Vol. 24, No. 2 (February 1996). The author states that 'present research suggests that with product differentiation may come significant increases in renewable demand', and that 'selling renewable technologies solely in the markets that want them, even assuming today's costs ("niche markets") could result in a dramatic shift … to renewables over the long term … based on a learning curve analysis, only niche markets need be captured in order to make significant gains in cost reductions and to create an orderly infrastructure build up. However, the potential market is much greater than the term niche would imply.'

thermal power development would occur first for local needs and might be closely related to developments in the gas industry, including field depletion.

Such developments would carry few of the systemic benefits discussed for embedded generation, and none of them is a near-term possibility, nor would they arise spontaneously. As long as gas remains abundant such renewables could be competitive with fossil fuels only if there was a significant externality charge (e.g. for carbon, or depletion as part of explicit diversity policy). Moreover, the scale of investment and integration required makes it likely that such developments could be achieved only in an explicit framework of European governmental cooperation.

Iceland has for many years claimed that it could export power at prices that are competitive on European markets, but it has never gained agreement for such a project from UK industries and governments. In Norway, the dominance of cheap hydro power – resulting in some of the lowest electricity prices in Europe – ruled out further hydro development or any serious investment in other power sources. It is only with the growing access to foreign markets, first through the Scandinavian system and now with the prospect of direct access to west European systems, that interest in additional generation is rising.

Access to markets has other financial consequences, because it allows international capital to be directed towards some of the most promising resources, wherever they are located. Two of the major obstacles to electricity development in many developing countries, including north Africa, have been capital shortage and low (often subsidized) electricity prices. Foreign investment, where allowed, could overcome the first of these but then faces the double obstacle of weak electricity prices and weak currencies: the investment requires hard currency but does not generate any.

The opening of European electricity markets opens up the possibility of European finance helping to construct solar power in north Africa and recouping some of the hard currency by electricity exports back to Europe. Obviously the economics are still problematic, given current technology, especially in competition with gas power generation in countries such as Algeria. The politics of generation for export could also be very difficult if the countries themselves are short of power. In some locations, however, the possibility could be opened up by the internationalization of European electricity.

The other problem is *integration*, in two senses. Integration of variable sources could be eased by interconnections. Internationalization of European electricity increases the regulating capacity of each system, since each can call upon the reserve of others; integration of Norwegian (and Alpine) hydro adds further to the regulating capacity. This obviously helps if large contributions from variable sources, such as large-scale development of North Sea wind energy, are considered.

The other and more important type of integration is structural integration of renewables with other sources and with infrastructure – for example, the combined use of different ocean-based renewables drawing upon existing or planned North Sea infrastructure, as discussed in Chapter 7. Until recently, such investment faced the institutional obstacle that the companies that developed and managed the infrastructure (e.g. oil rigs) or relevant manufacturing assets (e.g. shipyards) had no interest in electricity generation, and generating companies had no interest in using such infrastructure. The liberalization and internationalization of European electricity removes this obstacle. If an oil company or under-used shipyard sees an opportunity to construct a renewable energy system that, drawing on its existing investment, can compete with high-cost Danish or German electricity, for example, it will soon be able to do so. Again, the driving force is differentiation of electricity sources and markets that follows from liberalization: whoever is best placed to construct a renewable energy system, wherever it is best situated, will be able to do so in ways that were not possible in a centralized monopoly system. Renewable energy, because it is intrinsically more diverse than conventional sources, benefits more than conventional sources from such structures.

8.5 The paradox of scale: technical and political dimensions

Calling upon both localization and internationalization of electricity as favourable for renewables at first sight appears both technically contradictory and politically infeasible. Are they not opposite trends? Which takes precedence?

In fact, such development may not only be compatible but a natural reflection of technical and political developments. Technically, there is no reason why an electricity system should not take some of its electricity from local embedded resources and some from distant sources. Indeed, as part of the process of matching sources and loads in the most efficient manner, it is quite credible that some dispersed demands may be able to obtain power from local, flexible and quickly constructed renewable resources. But dispersed renewables have no advantage in meeting concentrated urban or industrial demands, and anyway on-land renewable resources are far from sufficient to meet all European energy needs. Conversely, there is no reason why concentrated demands should not obtain electricity from far afield; since the power has to be drawn from high-voltage transmission networks anyway, the added cost of transmission over long distances is minor, as are the transaction costs. The system-wide benefits of international integration, in terms of pooling and hence smoothing demand variations and drawing upon storage and regulating capacity, are well documented.

Institutionally, as argued above, liberalization enables localization provided that it is pursued far enough to allow decision-making on local electricity supplies to be made at the level of municipalities, rural councils, and even individual housing estates and buildings. In fact this is consistent with one of the strongest strands in the wider debate on sustainable development, which increasingly emphasizes 'local empowerment' as an important ingredient. The 'local Agenda 21' has proved one of the most important outputs of the Rio 'Earth Summit', and international alliances of local authorities regularly meet to discuss how local initiatives can contribute to sustainable development. The localization of European electricity would be well in keeping with this trend.

The initial stage of European electricity liberalization relates primarily to large industrial consumers; this is of no help in enabling localization, which will require liberalization to be pursued down to the level of municipalities and rural networks at least. Exploitation of PV for buildings, or of other renewables (including, for example, energy from wastes as considered in Volume III) for local DC networks or other systems, would require liberalization to proceed further still. Liberalization extending to individual users also enables dedicated companies to emerge offering 'green pricing', harnessing the preference of many people for renewable sources by offering them electricity generated by renewables at a premium price.

At the same time, international electricity trade is growing, as is the strength of international institutions, especially (but not exclusively) in the EU. The completion of the Single Market has necessitated greater regulatory oversight at the European level, and the same will be true for electricity: for a Scottish windfarm to compete on an equal footing with Austrian hydro in supplying power to German industrial consumers, for example, clearly requires some European oversight. The greater role of EU institutions in regulating the Single Market, combined with the basic political objectives of the European enterprise, has inevitably led to increased political strength of EU institutions *vis-à-vis* member states, notwithstanding the limitations of EU action with respect to energy policy discussed in Volume I.

From this perspective it is apparent that the likely evolution of electricity in Europe parallels the evolution of Europe's political structures: specifically, the much-hyped 'hollowing out' of the state will be matched by the 'hollowing out' of national electricity systems in favour of both localized generation and international trade. In both cases the process is at an early stage and it is difficult to predict how fast or how far it will proceed, but the trend is clear.

8.6 Policy tools

This book has argued that liberalization, despite the challenges it poses to current support regimes for renewables, is a prerequisite for obtaining large contributions from the dispersed primary renewables. To obtain the full potential, liberalization would need to be pursued down into the system, with cost-reflective pricing at different points in the network. This could give full rein to potential investors in local, small-scale and low-risk investments at the small-scale end, while allowing unimpeded international flows at the opposite end of the spectrum.

It should not be concluded from this that liberalization automatically favours renewable energy. I have argued that it is a *necessary* precondition for larger contributions from the primary renewables, but it is certainly not a *sufficient* one. Indeed, liberalization has at least three other general consequences that are inimical to most renewable sources.

- The first is the division and privatization of information. Although local agents (such as local authorities) may know most about local resources and constraints, they may themselves have little knowledge of the renewable energy technologies. Such information, being in many respects a common good, is notoriously undersupplied in competitive markets. No particular company can, for example, capture the benefits of educating architects in the use of PV or the benefits of RTD, and potential users may have little way of judging the quality and reliability of technologies proffered by different suppliers.
- Second, competitive markets generally create a focus upon short-term profits that is blind to long-run strategic issues. Yet it is these strategic issues that lie at the heart of sustainability and the long-run value of renewable sources, whereas the short-term focus discriminates disproportionately against capital-intensive investments such as the primary renewables.
- Third, competitive markets do not reflect the external costs and benefits of different energy sources, unless these are explicitly introduced by government policy. Yet introduction of such measures as energy- and carbon-taxation has proved extremely difficult, and even in countries that claim to have introduced them, the structure of such taxes is far removed from a realistic reflection of external costs.

Many of the policy mechanisms for overcoming these obstacles are generic to all renewables and are discussed in Volume IV of this study. Some, however, are specific to electricity systems. Information and corporate RTD has often been the first victim of privatization. The UK's former Central Electricity Generating Board, for example, led a number of important research efforts and appraisals, but all these efforts were curtailed, and the research laboratory was closed, in the aftermath of privatization. In the United States, electricity companies jointly fund the industry's Electric Power Research Institute to undertake research and analysis, and limited development, of industry-wide relevance. Especially as liberalization proceeds, industries in Europe would do well to establish a similar organization, perhaps as an expansion and recasting of the small existing Eurelectric organization that at present is primarily a Brussels-centred lobbying body.

To inject a longer-term focus for renewable energy investments, a number of governments already offer low-interest loans for renewable energy. Sometimes, as in the Netherlands, these are financed by a levy on electricity sales. Some forms of tax credits can have a similar effect. Such inducements to lengthen the time horizon of investments may be particularly important for the new primary renewable electricity sources, given the relatively high and capital-intensive investments involved, and the tendency of capital markets to increase further the rate of return sought, at least on the initial stages, given the perceived riskiness of relatively new technologies.

As the technologies develop and financiers become more confident, however, the question of whether the full value of renewable electricity is captured in the electricity price comes more to the fore. As argued at the beginning of this chapter, this value has internal *systemic* as well as *external* components. Liberalization, pursued appropriately, is a path towards realizing more of the systemic benefits. Yet at the same time it precludes any attempt to realize external benefits through explicit governmental direction of investments. As noted, attempts to use economic instruments to reflect the external costs of conventional sources are politically most problematic, because they affect the whole range of investments and the wholesale price of electricity.

Consequently, the dominant means to date towards supporting private renewable electricity generation has emerged in the form of various targeted supports that pay premium prices to renewable electricity generators, in ways that depend on the characteristics of the system. As long as the contribution of renewables remains fairly small, such targeted supports are politically easier than more generic measures such as carbon-taxation. But because they are specific and system-dependent, they become more visible and vulnerable as the renewable capacity expands, and/or as the structure of the system changes. Table 8.2 shows how expenditure on the UK's NFFO system expanded during the first half of the 1990s; already by 1995 it involved payments of about £100 million per year. Although this was still only one-tenth of the expenditure on supporting nuclear power stations, it nevertheless added about 1% to electricity bills, and the percentage is likely to rise substantially as the capacity expands. Formally, the NFFO policy was based upon the goal of price convergence with conventional sources, so

Table 8.2 Cost of the UK's renewable energy NFFO

Year	Total	Nuclear	Renewables	
	(£m)	(£m)	(£m)	(%)
1990–1	1,175	1,175	0	0
1991–2	1,324	1,311	13	1
1992–3	1,348	1,322	26	2
1993–4	1,234	1,166	68	5.5
1994–5	1,205	1,109	96	8

Source: Catherine Mitchell, 'A comparison of national models for the promotion of renewable energy resources', ALTENER contractors' meeting on power generation from renewables (September 1996).

that the NFFO could be terminated as a dedicated support mechanism; thus, it is intended more as a means of technology stimulation than as one reflecting external costs, and its extension to new rounds beyond 1998 is still uncertain at the time of writing.

In discussing options for the future support of renewable electricity in the UK, Mitchell[6] identifies three main mechanisms that can be targeted specifically at increasing payments for renewable electricity generation:

- *Standard payments:* This is by far the simplest mechanism, at least administratively. Utilities are required to purchase electricity from renewable sources, usually as a fixed percentage of the sale price. This is the primary system used in Germany and Denmark. It fixes the *price*, and any qualifying generator that can make a profit at the price offered may enter. As indicated in Table 1.6, in some countries the price varies according to the source.
- *NFFO bands:* This, the system used in the UK, is a bidding system in which potential renewable-energy generators bid for contracts, offering a price at which they could proceed. The government reviews the applications and selects a 'strike price', different for each technology band, based upon both price and quantity considerations. The regional electricity

---

[6] Catherine Mitchell, 'Future support of renewable energy in the UK – options and merits', *Energy and Environment*, Vol. 7, No. 3 (1996).

Table 8.3 Attributes of different renewable electricity support mechanisms

	Standard payment	Expanded NFFO bands	Banded renewables portfolio standard
Simplicity			
Administrative	Yes	No	No
Political	No	Yes	?
Known contract lengths	Yes	Yes	No
Known payment per kWh	Yes	Yes after order completed	No
Known market size	No	Yes after order completed	Yes
Who pays?			
Consumers	Yes	Yes	?
Suppliers	Yes	No	Yes
Balance for range of technologies including new entrants	Maybe	Yes	Yes
Good for energy traders, service cost	No	No	Yes
Complements planning system	Yes	?	Not if too ambitious
Time pressure?	Yes	No	No
Least-cost	No	No	Yes

Source: Adapted from C. Mitchell, 'Future support of renewable energy in the UK — options and merits', *Energy and Environment*, Vol. 7, No. 3 (1996).

companies pay this price to the generators, and are reimbursed for the extra cost (as compared with the electricity pool price) by the national fossil-fuel levy.

● *Renewables portfolio standard (RPS):* In this system, proposed in California and adopted in 1996 in a different form in the Netherlands, all electricity suppliers would be required to obtain renewable energy credits equivalent to some specified percentage of their annual sales.[7] These can be obtained by self-generation of renewables, by purchasing electricity directly from renewable-energy generators, or indirectly

---

[7] Mitchell refers to the system as 'Renewable Portfolio Standard', and the European Commission's Green Paper uses the term 'Renewable Energy Credits' (RECs). The latter could cause some confusion in the UK, where REC has been an acronym for Regional Electricity Company. Such 'credits' can be considered as the mechanisms for implementing portfolio standards.

through a renewable-energy trader. For a given time period the quantity is fixed irrespective of price, separately for different technologies if a banded system is used. This defines the targeted 'standard' portfolio for renewables generation.

Each approach has different problems and benefits, as summarized in Table 8.3 and discussed in more detail by Mitchell. She concludes that an expanded NFFO system would be the best option for continuing renewable energy supports in the UK. Here, the discussion focuses upon the long-run stability of such support systems and their appropriateness for the wide range of renewable energy technologies as their capacity increases and as national electricity systems become more competitive and internationally entwined.

From this perspective, each of these approaches faces distinct problems. Standard payments rely upon the *local electricity supplier* to finance the additional payments to renewables in its system, raising its average tariff. This means that systems with larger inputs from such renewables will be placed at a competitive disadvantage – hence the strident lobbying by the most affected German utilities against the standard payments under the Electricity Feed Law, and by Danish utilities against their standard payments system. Also, because it is a system that gives a direct payment rather than anything based on evaluation of competitive bids, the cost of the system is higher than the other approaches.

The NFFO system avoids competitiveness problems between different suppliers by averaging the cost of the support system across all electricity users in the UK, through the fossil-fuel levy. Nevertheless, it does raise the electricity price relative to countries that do not support renewable energy. Whereas this does not pose undue problems in the British context, such an approach may be more problematic for some continental countries that trade electricity more heavily, and like standard payments it limits the contribution to domestic resources, which may not be appropriate for countries such as the Netherlands with high population density and relatively limited renewable resources. Because payments are based on competitive bids, the cost tends to be less than under standard payments and the pressure for competitive reductions in generating costs is greater. However, evaluating bids is a relatively cumbersome and bureaucratic process that could become amplified as the number and size of bids grow.

Both NFFO and standard payment systems specify payments in terms of uniform prices for each technology band (albeit after bidding, for NFFO). This does nothing to encourage exploitation of the benefits of distributed generation: a generator would receive the same payment whether the output is correlated with demand or in opposition to it; and whether it contributes to meeting local demand at a weak point in the network or exacerbates system costs by generating electricity at the wrong times and surplus to local requirements. *Thus both these approaches render any moves towards either cost-reflective or 'green' pricing irrelevant, and neither encourages least-cost applications.* For the medium and long term this is an important drawback. It focuses investment upon the largest and least-cost generation, which might be centralized plants, rather than upon dispersed and relatively high-value applications such as PV façades on buildings. Indeed, very small applications such as PV façades would effectively be ruled out by the bureaucratic nature of the NFFO process; the transaction costs would exceed any likely returns.

The 'renewables portfolio standard' (RPS) system, like NFFO, would have to be banded to encourage investment in some of the less-developed renewables. Mitchell[8] points to various possible drawbacks associated with such a system, notably the possibility that if targets are set relatively high and soon, this could place undue pressure on the planning system by creating a rush for renewable capacity, perhaps with inadequate technology. Nevertheless, the drawbacks do not appear as important as the advantages. RPS is the only one of the three approaches that is consistent with the development of value-oriented, cost-reflective electricity pricing, thereby encouraging innovations in applications as well as technologies. This study has argued that such developments are an important part of the transition towards economically sustainable and large-scale development of renewables. Furthermore, by focusing on this aspect, the net cost of the system is minimized, which is not only valuable in its own right but also makes the support system less vulnerable to political objections based upon expense and adverse competitiveness impacts.

---

[8] Catherine Mitchell, 'Future support of renewable energy in the UK', op. cit.

Finally, an RPS-type system is the only one which could readily be internationalized, since if two or more countries created such a system the credits could be traded across European boundaries, even if the initial target percentages in different countries differed. With the growing internationalization of European electricity – and the possible development of ocean-based resources and others that might supply EU systems without being based there (such as Norwegian and Icelandic energy) – this is another important attribute. Fundamentally, such a system has the benefit of placing an obligation on electricity suppliers ultimately to fund renewable energy developments up to the specified amount, but separates this from the question of where the generation must take place, which may be an important consideration for diverse resources such as renewable energy. It can also be readily combined with other support mechanisms such as tax credits, extended demonstration schemes, etc. Hence it appears to be the most desirable approach for supporting renewables in the long run as the capacity and investment rise above the marginal.

Late in 1996, in its Green Paper on renewable energy, the European Commission proposed consideration of a system of 'Renewable Energy Credits', like the RPS approach outlined here, applied on a Europe-wide basis.[9] Optimistically, the Green Paper states that 'as it would apply to all utilities it would be competitively neutral'. In practice, however, since utilities have different renewable energy capacities, this would not necessarily be the case. Nevertheless, if the system were applied to sources other than large hydro, the capacities involved at present are so small that competitiveness impacts would be insignificant, particularly as most utilities in Europe are several years away from facing any serious competition. This does, however,

---

[9] 'Consideration could be given to the idea that a certain percentage of a Member State's electricity requirements will have to be met by renewables, enforced on each individual retail electricity supplier, with individual obligations tradeable through a system of Renewable Energy Credits ... [this] could serve a two-fold purpose if introduced on an EU-wide scale. Firstly it would promote renewables and secondly it would prevent market distortions arising from similar measures introduced by Member States ... [Utilities] would use technological applications which have the greatest value and they would use their resources and creativity to lower the cost of renewables.' European Commission, *Energy for the Future: Renewable Sources of Energy*, Green Paper for a Community Strategy, COM(96)576 (Brussels: 20 November 1996).

suggest that the longer the introduction of such a system is delayed, the more politically difficult it may become, as the capacity of renewable electricity sources grows differentially, and as competitive pressures increase.

Such a system can be considered either as a mechanism for supporting the commercialisation of emerging technologies, or as a means of reflecting external costs and benefits. In the former case, separate bands would be needed for different renewables to avoid crowding out some of the less developed technologies. Ideally other tools – such as input fuel and carbon taxes, or tradable emission permits – would be more efficient as a way of reflecting the external costs associated with conventional power sources. However, not only was the European carbon tax proposal essentially rejected, but even in the form developed by the Commission it applied to electricity supply, rather than generation inputs, with a rebate for renewable electricity generation – hardly an ideal tool either. This was in part because of the difficulties associated with accounting for electricity imports to the EU in the context of input-based resource/pollution taxes. In lieu of mechanisms for reflecting external costs directly in Europe's increasingly liberalised electricity systems, it is possible that tools such as RPS credits would also be established for the long term as a surrogate for reflecting some of the external benefits of renewable electricity. In this case, an unbanded RPS system could be considered alongside other programmes for promoting less developed renewable technologies.

The Green Paper implies that an RPS/credit system could be introduced uniformly across Europe. In fact, given the variation in system and resource conditions, expecting all member countries to adopt the same portfolio mix would not be fair, efficient or politically plausible. More likely is that countries would agree to set portfolio standards appropriate to their national conditions, perhaps in the context of EU negotiations on energy-environmental policy.

Indeed, shortly before going to press (March 1997), the EU Council reached an interim agreement on sharing CO_2 emission targets in the context of negotiations for the Kyoto Protocol on climate change, which is likely to result in legally binding commitments on CO_2 emissions over specified emission 'budget periods'. Renewable portfolio standard credits would accord with the structure of emission targets, and could be introduced as part of the package for implementing the Kyoto Protocol in Europe, with periods for the credits which would correspond to emission periods established by the Protocol.

8.7 Projections

What does all this imply for the future of European electricity and the role of primary renewables? First, it must be emphasized that systems in Europe still vary greatly in organization and structure. In particular, the French system, insulated from the thrust of European liberalization by the government's insistence on retaining the 'Single Buyer' option (Chapter 1), is unlikely to follow the general trend towards liberalization, at least for a very long time, and all the signs are that the French and probably Belgian systems will remain dominated by nuclear power (with some also exported throughout Europe). Elsewhere, the pace of liberalization and the way in which it is implemented may vary greatly, and renewable energy policies may remain correspondingly divergent at least in the near term while the capacity remains relatively small and the policies are not seriously challenged by the kinds of problems discussed in Section 8.5.

Nevertheless, the general trends implicit in the analysis of this book are likely to be widespread. Chapter 4 outlined three phases in the development of wind energy capacity:

- government-led and utility-led programmes, focusing upon demonstration and dissemination;
- early private-sector and utility investments under strong incentives that lead to an initially rapid expansion, but then run into constraints of public and/or utility opposition, land-use planning, and other obstacles associated with rapid industrial expansion;
- a more mature phase of somewhat slower growth (in percentage terms), based more upon local and national consensus-building and market-oriented supports.

Wind energy in Denmark has reached the third stage, and may be close to that in the UK, Germany, the Netherlands and perhaps Spain. The general model could be applicable to other renewables as well. Hydro power is in the third stage in most of Europe, though still with considerable scope for expansion of small hydro. The other primary renewables are still mostly in the first stage, and ocean- and desert-based systems are likely to remain so for a long time.

Hydro power is the most developed and familiar of the primary renewables, and the distributed benefits of small hydro are the most easily identified, since much of the resource is in remote regions; the capacity of small hydro can reasonably be expected at least to double over the next decade without major changes in the organization of European electricity systems or support mechanisms.

Wind energy is also relatively developed as a technology, but not as a generating industry. The interplay between impacts on the electricity system of this highly site-specific resource, the sensitivities surrounding siting and associated land-use planning, and the overall size and complexity of the resource mean that wind energy installation is likely to mature over a few decades, and the full use of offshore resources could extend this further. Installations in the 1990s have exceeded all projections, but long-term expansion may hinge upon more fundamental changes in systems and support mechanisms. Overall, this study has found no reason to dispute the broad picture set out by the European Wind Energy Association, of steady (and almost linear) expansion in the period 2000–2020 which could bring the contribution from wind energy to about 10% of European electricity.

PV is in theory a huge resource but its development will lag considerably behind wind and hydro energy. This is partly because it is growing from a still very small and high-cost base, but also because its effective exploitation requires quite radical changes in the organization of electricity systems, and a lengthy accumulation of experience concerning the best ways of managing and promoting building and other special siting applications. At the opposite end of the spectrum, ocean and desert-based renewables mostly still hinge upon development of technology and infrastructure; this also implies that their effective development may only mature over several decades.

This study has avoided the use of models in favour of less formal analysis, based upon the inherent characteristics of the resources, technologies and systems. Nevertheless, tentative estimates have been advanced of the

Table 8.4 Estimates of plausible contributions from different primary renewable electricity sources by 2030

Source	Contribution (%)[a]
Hydro	
Indigenous large	13—16
Indigenous small	3—5
Imports	1—5
Wind	
Onshore and near-shore	5—10
Offshore	1—5
Solar	
PV	2—8
Thermal power (mostly imports)	0—3
Ocean	
Tidal energy	< 1
Wave	1—3
Total	**26—56**

[a] For discussion and estimation see relevant chapters. The contributions are given as a percentage of total electricity demand, taking as an indicator 2500 TWh/yr consumption projected for the EU-15 in the year 2000. Actual realized percentage contributions could be scaled according to assumptions about changes in total electricity demand. However, more rapid demand growth might also be associated with more rapid development of some renewables and alleviation of some of the constraints that limit higher values in the percentage estimates, and with more pressure on fossil-fuel prices which would also aid the growth in renewables.

possible evolution and contributions over a period of two to three decades. These could result in contributions by 2030 in the range summarized in Table 8.4. Overall, this suggests that primary renewable electricity sources might on this timescale come to supply about 26–56% of the electricity demand projected for the turn of the century (about 2500 TWh).

It is hard to interpret this in terms of the percentage contribution to electricity supply in 2030, which depends upon long-term electricity demand as affected by prices, structural changes and efficiency policies; also, rapid demand growth would put more pressure upon fossil-fuel prices and alleviate some of the system constraints that influence the upper limits set out in

Table 8.4. Nevertheless (and particularly allowing for some additional contributions from the non-primary renewables, discussed in Volume III) we can conclude that renewable electricity sources are likely to contribute between a quarter and half of European electricity demand by 2030. Appropriate policies could lead towards the upper end of this range.

Two major sets of scenarios have explored the prospects for European electricity including renewables. The TERES II study, involving by far the most detailed modelling of renewable energy presented anywhere, contrasts a 'present policies' scenario against several different 'with policies' scenarios, projected out to 2020. It is forecast that with 'present policies' the primary renewables discussed in this volume would by 2020 contribute about 400 TWh, some 16% of likely EU-15 electricity generation in 2000, but most of this would be from already constructed large hydro schemes, the remainder being mostly small hydro (c.60 TWh) and wind (c.35 TWh).[10] In the 'with policies' scenarios:

- the contribution from small hydro expands only slightly, with a maximum of 85 TWh/yr;
- the contribution from wind energy more than doubles (to 90–120 TWh/yr by 2020);
- PV and ocean-based technologies contribute in only one scenario, with 'best practice' policies, and still supply under 20 and 30 TWh/yr respectively by 2020 even in this case.

Closer analysis of the scenarios reveals that the growth of renewable energy contributions slows down after 2010. This does not match to the conclusion of this study, which is that all the primary renewables except hydro are likely to expand at accelerating rates as costs continue to decline and as European electricity systems and policies evolve to accommodate them better. Probably, this aspect of the TERES result reflects the inherent limitations of such modelling studies, which can only poorly represent the

---

[10] Energy for Sustainable Development, *The European Renewable Energy Study II: The Prospects for Renewable Energy in 30 European Countries from 1995–2020*, EC DG-XVII (Brussels/Luxembourg: 1996). Contributions cited here are estimated from the graphs and converted from Mtoe at 11.63 TWh/Mtoe.

processes of technical evolution and structural change discussed in this book. In other words, and despite its unique sophistication, after 2010 the model used in the TERES projections may be running out of imagination rather than resources.

The same limitation is even more evident in the EC's main energy projections out to 2020. The Energy Directorate's study *European Energy to 2020*, summarized in Chapter 1, Section 1.4, projects that 450–500 GW of new electricity generating capacity will be installed over the period 1993–2020.[11] Although natural gas takes the largest share of new construction, 150–200 GW is projected to be new large-scale coal, oil or nuclear plant. Primary renewables are projected to account for 29 GW (of which 17 GW is hydro) in all scenarios except the more environmentally oriented 'Forum', which has 47 GW. The former would equate to about 30–50 TWh/yr from non-hydro renewables; the latter to about 100–130 TWh/yr.[12]

Such projections are hard to justify on the basis of the analysis set out in this book, or the TERES II study. If and as liberalization proceeds, particularly if it is in tandem with environmental and CO_2 pressures, it is hard to see who will invest in 150–200 GW of new, centralized nuclear, coal and oil capacity. Similarly, the renewable energy contribution seems implausibly small in all except the 'Forum' scenario, and even this contribution is no bigger than the projected output from wind energy alone in the TERES with-policies scenarios.

Nevertheless, it is clear that the development of renewables is highly dependent upon policies. The underlying policy question is how governments in Europe will implement the progressive liberalization of electricity (and gas) systems, and what attitudes they will take to its consequences. Specifically, in its simplest form – that of allowing Open Access at the level of industries and municipalities – it is likely to have one major consequence for generation: accelerating the move towards natural gas in Europe's power systems. Technically, natural gas and renewables are highly compatible, and

---

[11] Energy in Europe Special Issue, *European Energy to 2020: A Scenario Approach*, DG-XVII (Brussels/Luxembourg: Spring 1996).
[12] Generation data are not given; these figures are converted from GW assuming load factor of 30–50%, based on an assumed large share of wind energy.

together could form the basic components of a very clean electricity system. But economically, on the raw criteria used by private-sector investors in a competitive market, liberalization is likely to result only in gas, gas and more gas.

Section 8.6 discussed targeted policies for promoting market expansion of renewable electricity sources in competitive systems, and in Section 8.2 a dozen potential reasons for employing such mechanisms were sketched out. The rapid development of renewable energy technologies, and the prospects for further cost reductions as markets and industries expand, mean that the cost of such support measures is unlikely to be very large if they are well designed; nevertheless, to support a non-hydro renewable energy contribution of 10% of EU electricity generation could involve implicit transfers of several billion ecus.[13] More likely, European structures would evolve over time away from reliance on direct output-credit supports for the more mature renewables, and more towards systems of cost-reflective pricing including externality charges in some form that would start to reflect better the 'holistic' economics of renewables discussed at the beginning of this chapter.

8.8 Prospects

Given such policies, the analysis of this book suggests a rather different outlook from any sketched in the studies mentioned above. Natural-gas plants are likely to dominate electricity investments over the next couple of decades at least, but if governments continued to pursue policies to foster the growth of renewable electricity sources, there would be a progressive evolution and expansion of their contribution according to technology and system. Of the primary renewables, small hydro would be the first beneficiary, but its resource is limited; the wind energy contribution would follow on its heels and ultimately overtake it. As liberalization digs deeper

---

[13] If policies are well designed, and given inevitable cost reductions with expanding markets, the average cost differential with natural gas is unlikely to exceed 1–2 ∈/kWh. Even so, at this average differential, for renewables to supply 250 TWh (10% of EU 2000 generation) would require 2.5–5 billion ecus.

into European electricity systems, PV applications would expand, but the limitations of the building-based 'niche' markets suggest that its niche contributions could only exceed wind energy if there were as yet unforeseen developments.

Besides these relatively limited niche applications, both PV and ocean-based energy sources are likely to remain substantially more expensive than natural gas or oil for as long as Europe has access to adequate reserves of these fuels, and maybe than other conventional sources. But if a return to coal is ruled out by climatic concerns, as seems likely, the basis for later expansion could be set by development of PV at the small end of the scale, and at the large-scale end by the development of technologies and infrastructure for ocean-based renewables in northern Europe, including Norwegian and Icelandic resources, and for desert-based renewables in north Africa. PV might penetrate into transport markets if and as air quality problems and longer-term concerns about oil supply lead towards electric vehicles. Electricity from non-fossil sources (perhaps including nuclear power) transmitted over longer distances could assume a greater role, particularly if and as concerns rise about dependence upon increasingly concentrated and limited gas resources and pipelines. Such long-term resources could comprise a mixture of ocean-based renewables, solar thermal power from African deserts, and large-scale hydro from Africa or Russia.

Combined with structural trends including wider use of co-generation (combined heat and power) and the biomass and heat-based resources considered in Volume III, all this sets the stage for the evolution of European electricity towards an intrinsically cleaner and more efficient system over time. It could evolve over a period of some decades, through peak use of natural gas towards a system drawing heavily upon the distributed use of renewables for supplying dispersed demands, and growing long-distance transmission of other non-fossil sources to supply concentrated industrial and some city demands. It is not possible to predict either the exact mix or the timing of the different stages of such an evolution, but the general prospect is one that could – and should – entice policy-makers for decades to come.